STUDENT SOLUTIONS MANUAL

Charles H. Corwin
American River College

Donald D. Lucas
American River College

# Chemistry
## Concepts and Connections

Charles H. Corwin
American River College

Prentice Hall
Englewood Cliffs, New Jersey 07632

Editorial/production supervision: ***Joanne E. Jimenez***
Acquisitions editor: ***Tim Bozik***
Supplements acquisitions editor: ***Mary Hornby***
Prepress buyer: ***Paula Massenaro***
Manufacturing buyer: ***Lori Bulwin***

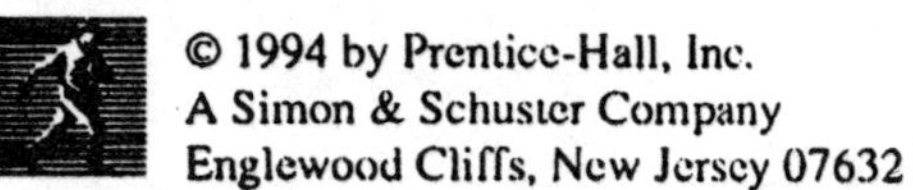


A Simon & Schuster Company
Englewood Cliffs, New Jersey 07632



Printed in the United States of America

10 9 8 7 6 5

ISBN 0-13-482357-5

PRENTICE-HALL INTERNATIONAL (UK) LIMITED, LONDON
PRENTICE-HALL OF AUSTRALIA PTY. LIMITED, SYDNEY
PRENTICE-HALL CANADA INC. TORONTO
PRENTICE-HALL HISPANOAMERICANA, S.A., MEXICO
PRENTICE-HALL OF INDIA PRIVATE LIMITED, NEW DELHI
PRENTICE-HALL OF JAPAN, INC., TOKYO
SIMON & SCHUSTER ASIA PTE. LTD., SINGAPORE
EDITORA PRENTICE-HALL DO BRASIL, LTDA., RIO DE JANEIRO

SOLUTIONS

# Contents

SOLUTIONS

# Preface

***To the Student***

The overall objective of this *Student Solutions Manual* that accompanies *Chemistry: Concepts and Connections* is to provide detailed solutions for the odd-numbered exercises at the end of each chapter. In addition to providing solutions, this solutions manual has built-in features to make it "student friendly." These features include:

- a consistent, stepwise presentation of the final answers in an easy to follow table format.
- a complete solution using the *unit analysis method of problem solving* is provided for all but a few calculations.
- easy-to-follow cancellation of units for each unit analysis solution.
- all stoichiometry-related problems begin with a corresponding balanced chemical equation.
- computer-generated graphics to help visualize problems involving graphs, orbitals, electron dot formulas, and electrochemical cells.

The *Student Solutions Manual* can be a powerful study aid and can accelerate the learning process if used appropriately. The following suggestions should help guide you in a positive direction.

- Refer to the solutions manual only after you have attempted a problem but have been unsuccessful in making progress.
- If you become "stuck" on a problem, glance quickly at the solution only to gain insight about the correct problem-solving procedure.
- After a problem has been solved, compare your solution to that in the solutions manual using a step-by-step fashion; make sure you understand each step of the solution and not just the answer.

We ask that you follow these suggestions and wish you success in your chemistry class and may you enjoy your experience. Lastly, we hope that the *concepts* you learn will help you make *connections* with this fascinating subject.

In the spirit of cooperative learning, we hope this solutions manual makes your job of learning chemistry a little easier. For those questions that we have not answered for you, visit your professor during office hours. Most professors are only too happy to help students, especially students who are making a serious effort to complete assignments and learn the material. As a further study aid, your professor may choose to share the detailed solutions to the even-numbered exercises that appear in the *Instructor's Resource Manual.*

**Acknowledgments**

This *Student Solutions Manual** represents a collaborative effort by an experienced chemistry tutor and by an introductory chemistry teacher who have listened carefully to hundreds of students to better understand the areas they find most troublesome. Together, we would like to acknowledge the excellent facilities and staff at the American River College Learning Resource Center which encourage students to come and ask frank questions that expose the heart of their difficulty.

D.D.L., in particular, would like to thank the following professors who were responsible either directly or indirectly for the quality of this solutions manual: John Newey, Nancy Reitz, Stephen Ruis, and Michael Scott.

*Donald D. Lucas*
*Learning Resource Center*
*American River College*
*Sacramento, CA 95841*

*Charles H. Corwin*
*Department of Chemistry*
*American River College*
*Sacramento, CA 95841*

* This document was created on a Macintosh™ Classic II, using Microsoft® Word 5.0, and printed on a Personal LaserWriter™ NT with Palatino, Helvetica, and New Century Schoolbook fonts.

# A Mathematical Foundation

# CHAPTER 1

## Section 1.1 *Uncertainty in Measurement*

1. No measurements are exact because all instruments have uncertainty.

3.

| | Instrument | Quantity | | Instrument | Quantity |
|---|---|---|---|---|---|
| (a) | metric ruler | length | (b) | buret | volume |
| (c) | balance | mass | (d) | pipet | volume |
| (e) | stopwatch | time | (f) | grad. cylinder | volume |

5.

| | Maximum | Minimum | | Maximum | Minimum |
|---|---|---|---|---|---|
| (a) | 6.6 cm | 6.4 cm | (b) | 0.52 g | 0.50 g |
| (c) | 10.1 mL | 9.9 mL | (d) | 35.6 s | 35.4 s |

## Section 1.2 *Significant Digits*

7.

| | | |
|---|---|---|
| (a) | 0.05 cm | 1 significant digit |
| (b) | 12.0 cm | 3 significant digits |
| (c) | 0.707 g | 3 significant digits |
| (d) | 1.110 g | 4 significant digits |
| (e) | 24.60 mL | 4 significant digits |
| (f) | 100 mL | 1 significant digit |
| (g) | $3.71 \times 10^3$ s | 3 significant digits |
| (h) | $1.000 \times 10^{-1}$ s | 4 significant digits |

## Section 1.3 *Rounding Off Nonsignificant Digits*

9.

| | Not Rounded | Rounded |
|---|---|---|
| (a) | 31.505 | 31.5 |
| (b) | 213,600 | 214,000 |
| (c) | 5.155 | 5.16 |
| (d) | 77.504 | 77.5 |
| (e) | 0.01842 | 0.0184 |
| (f) | 0.000 000 484 500 | 0.000 000 485 |
| (g) | $2.571 \times 10^5$ | $2.57 \times 10^5$ |
| (h) | $5.6954 \times 10^{-2}$ | $5.70 \times 10^{-2}$ |

## Section 1.4 *Adding and Subtracting Measurements*

11.

| | | Answer |
|---|---|---|
| (a) | 31.15 cm + 0.5 cm = 31.65 cm | 31.7 cm |
| (b) | 50.2 cm + 5.25 cm = 55.45 cm | 55.5 cm |
| (c) | 0.4 g + 0.44 g + 0.444 g = 1.284 g | 1.3 g |
| (d) | 15.5 g + 7.50 g + 0.050 g = 23.050 g | 23.1 g |
| (e) | 0.55 mL + 36.15 mL + 17.3 mL = 54.00 mL | 54.0 mL |
| (f) | 4 mL + 16.3 mL + 0.95 mL = 21.25 mL | 21 mL |

## Section 1.5 *Multiplying and Dividing Measurements*

13.

| | | Answer |
|---|---|---|
| (a) | 3.65 cm × 2.10 cm = 7.665 $cm^2$ | 7.67 $cm^2$ |
| (b) | 8.75 cm × 1.15 cm = 10.0625 $cm^2$ | 10.1 $cm^2$ |
| (c) | 16.5 cm × 1.7 cm = 28.05 $cm^2$ | 28 $cm^2$ |
| (d) | 21.1 cm × 20 cm = 422 $cm^2$ | 400 $cm^2$ |
| (e) | 5.1 cm × 1.25 cm × 0.5 cm = 3.1875 $cm^3$ | 3 $cm^3$ |
| (f) | 5.15 cm × 2.55 cm × 1.1 cm = 14.44575 $cm^3$ | 14 $cm^3$ |
| (g) | 12.0 $cm^2$ × 1.00 cm = 12.0 $cm^3$ | 12.0 $cm^3$ |
| (h) | 22.1 $cm^2$ × 0.75 cm = 16.575 $cm^3$ | 17 $cm^3$ |

## Section 1.6 *Exponential Number*

15. (a) $3 \times 3 = 3^2$ (b) $2 \times 2 \times 2 \times 2 \times 2 \times 2 = 2^6$

(c) $\frac{1}{2} \times \frac{1}{2} \times \frac{1}{2} = 2^{-3}$ (d) $\frac{1}{3} \times \frac{1}{3} \times \frac{1}{3} \times \frac{1}{3} \times \frac{1}{3} = 3^{-5}$

(e) $10 \times 10 \times 10 = 10^3$ (f) $10 \times 10 \times 10 \times 10 \times 10 \times 10 = 10^6$

(g) $\frac{1}{10} \times \frac{1}{10} \times \frac{1}{10} = 10^{-3}$ (h) $\frac{1}{10} \times \frac{1}{10} \times \frac{1}{10} \times \frac{1}{10} = 10^{-4}$

17.

| | Number | Power of Ten |
|---|---|---|
| (a) | 1 | $1 \times 10^0$ |
| (b) | 0.1 | $1 \times 10^{-1}$ |
| (c) | 1,000,000 | $1 \times 10^6$ |
| (d) | 0.0001 | $1 \times 10^{-4}$ |
| (e) | 10,000,000,000 | $1 \times 10^{10}$ |
| (f) | 0.000 001 | $1 \times 10^{-6}$ |
| (g) | 100,000,000,000,000,000 | $1 \times 10^{17}$ |
| (h) | 0.000 000 000 000 001 | $1 \times 10^{-15}$ |

**Section 1.7** *Scientific Notation*

19.

| | Number | Scientific Notation |
|---|---|---|
| (a) | 80,916,000 | $8.0916 \times 10^{7}$ |
| (b) | 0.000 000 015 | $1.5 \times 10^{-8}$ |
| (c) | 335,600,000,000,000 | $3.356 \times 10^{14}$ |
| (d) | 0.000 000 000 000 927 | $9.27 \times 10^{-13}$ |

21.

| Number | Scientific Notation |
|---|---|
| 26,900,000,000,000,000,000,000 neon atoms | $2.69 \times 10^{22}$ neon atoms |

**Section 1.8** *Calculators and Significant Digits*

23.

| | | | Calculator Notation | | Scientific Notation |
|---|---|---|---|---|---|
| (a) | $(4.65 \times 10^{5})(9.5 \times 10^{2})$ | = | 4.4 E08 | = | $4.4 \times 10^{8}$ |
| (b) | $\dfrac{(3.62 \times 10^{4})}{(1.60 \times 10^{-3})}$ | = | 2.26 E07 | = | $2.26 \times 10^{7}$ |
| (c) | $(2.33 \times 10^{5})(11.21 \times 10^{3})$ | = | 2.61 E09 | = | $2.61 \times 10^{9}$ |
| (d) | $\dfrac{(1.11 \times 10^{8})}{(2.05 \times 10^{-11})}$ | = | 5.41 E18 | = | $5.41 \times 10^{18}$ |
| (e) | $(5.97 \times 10^{-3})(81.4 \times 10^{-6})$ | = | 4.86 E–07 | = | $4.86 \times 10^{-7}$ |
| (f) | $\dfrac{(54.9 \times 10^{6})}{(7.87 \times 10^{4})}$ | = | 6.98 E02 | = | $6.98 \times 10^{2}$ |
| (g) | $(1.16 \times 10^{4})(3.14 \times 10^{-4})$ | = | 3.64 E00 | = | $3.64 \times 10^{0}$ |
| (h) | $\dfrac{(11.17 \times 10^{-4})}{(25.3 \times 10^{-5})}$ | = | 4.42 E00 | = | $4.42 \times 10^{0}$ |
| (i) | $(8.68 \times 10^{-1})(6.13 \times 10^{-2})$ | = | 5.32 E–02 | = | $5.32 \times 10^{-2}$ |
| (j) | $\dfrac{(0.176 \times 10^{11})}{(4.58 \times 10^{-19})}$ | = | 3.84 E28 | = | $3.84 \times 10^{28}$ |

## **Section 1.9** *Unit Conversion Factors*

25.

| | Unit Equation | Unit Factors | | |
|---|---|---|---|---|
| (a) | 1 dollar = 10 dimes | $\frac{\text{1 dollar}}{\text{10 dimes}}$ | and | $\frac{\text{10 dimes}}{\text{1 dollar}}$ |
| (b) | 1 nickel = 5 pennies | $\frac{\text{1 nickel}}{\text{5 pennies}}$ | and | $\frac{\text{5 pennies}}{\text{1 nickel}}$ |
| (c) | 1 dollar = 4 quarters | $\frac{\text{1 dollar}}{\text{4 quarters}}$ | and | $\frac{\text{4 quarters}}{\text{1 dollar}}$ |
| (d) | 10 pennies = 1 dime | $\frac{\text{10 pennies}}{\text{1 dime}}$ | and | $\frac{\text{1 dime}}{\text{10 pennies}}$ |
| (e) | 2 nickels = 1 dime | $\frac{\text{2 nickels}}{\text{1 dime}}$ | and | $\frac{\text{1 dime}}{\text{2 nickels}}$ |
| (f) | 1 half-dollar = 10 nickels | $\frac{\text{1 half-dollar}}{\text{10 nickels}}$ | and | $\frac{\text{10 nickels}}{\text{1 half-dollar}}$ |
| (g) | 50 pennies = 1 half-dollar | $\frac{\text{50 pennies}}{\text{1 half-dollar}}$ | and | $\frac{\text{1 half-dollar}}{\text{50 pennies}}$ |
| (h) | 25 pennies = 1 quarter | $\frac{\text{25 pennies}}{\text{1 quarter}}$ | and | $\frac{\text{1 quarter}}{\text{25 pennies}}$ |
| (i) | 1 quarter = 5 nickels | $\frac{\text{1 quarter}}{\text{5 nickels}}$ | and | $\frac{\text{5 nickels}}{\text{1 quarter}}$ |
| (j) | 5 dimes = 1 half-dollar | $\frac{\text{5 dimes}}{\text{1 half-dollar}}$ | and | $\frac{\text{1 half-dollar}}{\text{5 dimes}}$ |

## **Section 1.10** *Problem Solving by Unit Analysis*

27. (a) $2 \cancel{\text{dollars}} \times \frac{\text{20 nickels}}{1 \cancel{\text{dollar}}} = \text{40 nickels}$

(b) $2150 \cancel{\text{dimes}} \times \frac{\text{1 dollar}}{10 \cancel{\text{dimes}}} = \text{215 dollars}$

(c) $0.750 \cancel{\text{gross}} \times \frac{\text{144 pencils}}{1 \cancel{\text{gross}}} = \text{108 pencils}$

(d) $12{,}500 \cancel{\text{sheets}} \times \frac{\text{1 ream}}{500 \cancel{\text{sheets}}} = \text{25 reams}$

(e) $75.0 \cancel{\text{ounces}} \times \frac{\$6.50}{1 \cancel{\text{ounce}}} = \$487.50$

(f) $100.0 \cancel{\text{ounces}} \times \frac{\$425.50}{1 \cancel{\text{ounce}}} = \$42{,}550.00$

**Section 1.11** *Percentage*

29. (a) $\frac{101 \text{ students}}{5846 \text{ students}} \times 100 = 1.73\%$ are chemistry majors

(b) O+: $\frac{21{,}594 \text{ patients}}{55{,}368 \text{ patients}} \times 100 = 39.0\%$ are O+ type

A+: $\frac{18{,}825 \text{ patients}}{55{,}368 \text{ patients}} \times 100 = 34.0\%$ are A+ type

B+: $\frac{4706 \text{ patients}}{55{,}368 \text{ patients}} \times 100 = 8.50\%$ are B+ type

AB+: $\frac{1938 \text{ patients}}{55{,}368 \text{ patients}} \times 100 = 3.50\%$ are AB+ type

O–: $\frac{3876 \text{ patients}}{55{,}368 \text{ patients}} \times 100 = 7.00\%$ are O– type

A–: $\frac{3322 \text{ patients}}{55{,}368 \text{ patients}} \times 100 = 6.00\%$ are A– type

B–: $\frac{831 \text{ patients}}{55{,}368 \text{ patients}} \times 100 = 1.50\%$ are B–type

AB–: $\frac{276 \text{ patients}}{55{,}368 \text{ patients}} \times 100 = 0.50\%$ are AB– type

(c) $\frac{34.1 \text{ g aluminum}}{100 \text{ g bauxite}} \times 5.750 \text{ g bauxite} = 1.96$ g aluminum

(d) $\frac{100 \text{ mL air}}{20.9 \text{ mL oxygen}} \times 225 \text{ mL oxygen} = 1080$ mL air

(e) all quantities: 250 mL ethyl alcohol + 375 mL water = 625 mL solution

$\frac{250 \text{ mL}}{625 \text{ mL}} \times 100 = 40.0\%$ of solution is ethyl alcohol

(f) $\frac{1.50 \text{ gallons}}{12.5 \text{ gallons}} \times 100 = 12.0\%$ of gasohol is ethanol

31. Total quantity = 3.051 g copper and zinc
Quantity of copper = 2.898 g
Quantity of zinc = total quantity – quantity of copper
Quantity of zinc = 3.051 g – 2.898 g = 0.153 g

Percent of copper: $\frac{2.898 \text{ g}}{3.051 \text{ g}} \times 100 = 94.99\%$ copper

Percent of zinc: $\frac{0.153 \text{ g}}{3.051 \text{ g}} \times 100 = 5.01\%$ zinc

## General Exercises

33. Ruler A: 1.5 cm (± 0.1 cm)
Ruler B: 1.45 cm (± 0.05 cm)

35. 10.0 mL

37.

| Not Rounded | Rounded |
|---|---|
| 22.989768 atomic mass units | 23.0 atomic mass units |

39. 126.457 g + 131.6 g = 258.057 g (not rounded) Total mass = 258.1 g

41.

| | Exponential Number | Scientific Notation |
|---|---|---|
| (a) | $352 \times 10^4$ | $3.52 \times 10^6$ |
| (b) | $416 \times 10^3$ | $4.16 \times 10^5$ |
| (c) | $0.170 \times 10^2$ | $1.70 \times 10^1$ |
| (d) | $0.00125 \times 10^2$ | $1.25 \times 10^{-1}$ |

43.

| Scientific Notation | Numerical Form |
|---|---|
| $3.20 \times 10^{20}$ | 320,000,000,000,000,000,000 |

45. Total mass: $9.10953 \times 10^{-28}$ g + $1.67265 \times 10^{-24}$ g = $1.67356 \times 10^{-24}$ g

47. one metric ton = 2200 pounds
one English ton = 2000 pounds
2200 pounds – 2000 pounds = 200 pounds

49. Volume = length × length × length
length = $4.32 \times 10^{-1}$ cm = 0.432 cm
Volume = 0.432 cm × 0.432 cm × 0.432 cm = 0.0806 $cm^3$

51. 4.0 billion years = $4.0 \times 10^9$ years

$$4.0 \times 10^9 \text{ \sout{years}} \times \frac{365 \text{ \sout{days}}}{1 \text{ \sout{year}}} \times \frac{24 \text{ hours}}{1 \text{ \sout{day}}} = 3.5 \times 10^{13} \text{ hours}$$

53. $$1 \text{ \sout{troy pound}} \times \frac{12 \text{ \sout{troy ounces}}}{1 \text{ \sout{troy pound}}} \times \frac{31.1 \text{ grams}}{1 \text{ \sout{troy ounce}}} = 373 \text{ grams of gold}$$

55. 1 troy pound = 373 grams of gold
1 avoirdupois pound = 454 grams of feathers

Since 373 grams of gold is less than 454 grams of feathers, a pound of feathers (avoirdupois) weighs more than a pound of gold (troy).

# Measurement

CHAPTER 2

## Section 2.1 *The Metric System*

1. All of the given statements regarding the metric system are true.

3.

| | Quantity | Metric Unit | Symbol |
|---|---|---|---|
| (a) | length | meter | m |
| (b) | time | second | s |
| (c) | mass | kilogram | kg |
| (d) | temperature | degree Celsius | °C |
| (e) | volume | liter | L |
| (f) | heat | calorie | cal |

5.

| | Metric Unit | Symbol | | Metric Unit | Symbol |
|---|---|---|---|---|---|
| (a) | microgram | μg | (b) | micrometer | μm |
| (c) | kilometer | km | (d) | millisecond | ms |
| (e) | deciliter | dL | (f) | kilocalorie | kcal |
| (g) | centiliter | cL | (h) | cubic centimeter | $cm^3$ |
| (i) | nanogram | ng | (j) | square meter | $m^2$ |

## Section 2.2 *Metric Unit Conversion Factors*

7.

| | Unit Equation | | Unit Equation |
|---|---|---|---|
| (a) | 1000 mm = 1 m | (b) | 10 dL = 1 L |
| (c) | 1 g = $10^6$ μg | (d) | 10 dm = 1 m |
| (e) | 1 kL = 1000 L | (f) | $10^9$ ns = 1 s |

9.

| | Unit Equation | | Unit Equation |
|---|---|---|---|
| (a) | 1 m = $10^9$ nm | (b) | $10^6$ μm = 1 m |
| (c) | 100 cg = 1 g | (d) | 1 mL = 1 $cm^3$ |
| (e) | 1 L = 10 dL | (f) | 1000 cal = 1 kcal |

## Section 2.3 *Metric Problems by Unit Analysis*

11. Step 1: Write down the units of the unknown.
Step 2: Write down the relevant given value.
Step 3: Apply one or more unit factors.

**Section 2.4** *Metric–Metric Unit Conversions*

13. (a) $1.55\ \cancel{\text{km}} \times \dfrac{1000\ \text{m}}{1\ \cancel{\text{km}}} = 1550\ \text{m}$

(b) $10.6\ \cancel{\text{dg}} \times \dfrac{1\ \text{g}}{10\ \cancel{\text{dg}}} = 1.06\ \text{g}$

(c) $0.486\ \cancel{\text{g}} \times \dfrac{100\ \text{cg}}{1\ \cancel{\text{g}}} = 48.6\ \text{cg}$

(d) $125\ \cancel{\text{mL}} \times \dfrac{1\ \text{L}}{1000\ \cancel{\text{mL}}} = 0.125\ \text{L}$

(e) $1.885\ \cancel{\text{L}} \times \dfrac{10\ \text{dL}}{1\ \cancel{\text{L}}} = 18.85\ \text{dL}$

(f) $100\ \cancel{\text{ns}} \times \dfrac{1\ \text{s}}{1 \times 10^{9}\ \cancel{\text{ns}}} = 1 \times 10^{-7}\ \text{s}$

(g) $0.388\ \cancel{\text{m}} \times \dfrac{1000\ \text{mm}}{1\ \cancel{\text{m}}} = 388\ \text{mm}$

(h) $94.6\ \cancel{\text{kcal}} \times \dfrac{1000\ \text{cal}}{1\ \cancel{\text{kcal}}} = 94{,}600\ \text{cal}\ \ (9.46 \times 10^{4}\ \text{cal})$

(i) $0.000125\ \cancel{\text{s}} \times \dfrac{1000\ \text{ms}}{1\ \cancel{\text{s}}} = 0.125\ \text{ms}$

(j) $3.15 \times 10^{5}\ \cancel{\text{mg}} \times \dfrac{1\ \text{g}}{1000\ \cancel{\text{mg}}} = 3.15 \times 10^{2}\ \text{g}\ \ (315\ \text{g})$

**Section 2.5** *Metric–English Unit Conversions*

15. (a) 2.54 cm = 1 in. (b) 454 g = 1 lb
(c) 946 mL = 1 qt (d) $946\ \text{cm}^3 = 1\ \text{qt}$

17. (a) $66\ \cancel{\text{in.}} \times \dfrac{2.54\ \text{cm}}{1\ \cancel{\text{in.}}} = 170\ \text{cm}$

(b) $86\ \cancel{\text{cm}} \times \dfrac{1\ \text{in.}}{2.54\ \cancel{\text{cm}}} = 34\ \text{in.}$

(c) $1.01\ \cancel{\text{lb}} \times \dfrac{454\ \text{g}}{1\ \cancel{\text{lb}}} = 459\ \text{g}$

(d) $36\ \cancel{\text{g}} \times \dfrac{1\ \text{lb}}{454\ \cancel{\text{g}}} = 0.079\ \text{lb}$

(e) $0.500\ \cancel{\text{qt}} \times \dfrac{946\ \text{mL}}{1\ \cancel{\text{qt}}} = 473\ \text{mL}$

(f) $750\ \cancel{\text{mL}} \times \dfrac{1\ \text{qt}}{946\ \cancel{\text{mL}}} = 0.79\ \text{qt}$

19. $\dfrac{52\ \cancel{\text{miles}}}{1\ \cancel{\text{gallon}}} \times \dfrac{1\ \cancel{\text{gallon}}}{4\ \cancel{\text{qt}}} \times \dfrac{1\ \cancel{\text{qt}}}{946\ \cancel{\text{mL}}} \times \dfrac{1000\ \cancel{\text{mL}}}{1\ \text{L}} \times \dfrac{1.61\ \text{km}}{1\ \cancel{\text{mile}}} = 22\ \text{km/L}$

21. $\dfrac{1200\ \cancel{\text{ft}}}{1\ \text{s}} \times \dfrac{12\ \cancel{\text{in.}}}{1\ \cancel{\text{ft}}} \times \dfrac{2.54\ \cancel{\text{cm}}}{1\ \cancel{\text{in.}}} \times \dfrac{1\ \text{m}}{100\ \cancel{\text{cm}}} = 370\ \text{m/s}$

**Section 2.6** *Volume by Calculation*

23. $5.08\ \text{cm} \times 10.2\ \text{cm} \times 3.05\ \cancel{\text{m}} \times \dfrac{100\ \text{cm}}{1\ \cancel{\text{m}}} = 15{,}800\ \text{cm}^3$

25. $\dfrac{(15.3\ \text{cm}^3)}{(4.95\ \text{cm})\,(2.45\ \text{cm})} = 1.26\ \text{cm}$

27. (a) $1\ \cancel{\text{L}} \times \dfrac{1000\ \text{mL}}{\cancel{\text{L}}} = 1000\ \text{mL}$

(b) $1\ \cancel{\text{L}} \times \dfrac{1000\ \cancel{\text{mL}}}{\cancel{\text{L}}} \times \dfrac{1\ \text{cm}^3}{1\ \cancel{\text{mL}}} = 1000\ \text{cm}^3$

(c) $1\ \cancel{\text{cm}}^3 \times \dfrac{1\ \text{mL}}{\cancel{\text{cm}}^3} = 1\ \text{mL}$

(d) $1\ \cancel{\text{in.}}^3 \times \dfrac{2.54\ \text{cm}}{1\ \cancel{\text{in.}}} \times \dfrac{2.54\ \text{cm}}{1\ \cancel{\text{in.}}} \times \dfrac{2.54\ \text{cm}}{1\ \cancel{\text{in.}}} = 16.4\ \text{cm}^3$

29. $415\ \cancel{\text{in.}}^3 \times \dfrac{2.54\ \cancel{\text{cm}}}{1\ \cancel{\text{in.}}} \times \dfrac{2.54\ \cancel{\text{cm}}}{1\ \cancel{\text{in.}}} \times \dfrac{2.54\ \cancel{\text{cm}}}{1\ \cancel{\text{in.}}} \times \dfrac{1\ \cancel{\text{mL}}}{1\ \cancel{\text{cm}}^3} \times \dfrac{1\ \text{L}}{1000\ \cancel{\text{mL}}} =$
$6.80\ \text{L}$

**Section 2.7** *Volume by Displacement*

31. volume of iron pyrite: 55.0 mL – 44.5 mL = 10.5 mL

**Section 2.8** *Density*

33.

| | Substance | Sink or Float | | Substance | Sink or Float |
|---|---|---|---|---|---|
| (a) | kerosene | float | (b) | paraffin wax | float |
| (c) | olive oil | float | (d) | birch wood | float |
| (e) | chloroform | sink | (f) | limestone | sink |

35. (a) $250\ \cancel{\text{mL}} \times \dfrac{0.69\ \text{g}}{\cancel{\text{mL}}} = 170\ \text{g}$

(b) $36.5\ \cancel{\text{mL}} \times \dfrac{0.791\ \text{g}}{\cancel{\text{mL}}} = 28.9\ \text{g}$

(c) $0.75\ \cancel{\text{cm}}^3 \times \dfrac{2.18\ \text{g}}{\cancel{\text{cm}}^3} = 1.6\ \text{g}$

(d) $455\ \cancel{\text{cm}}^3 \times \dfrac{1.715\ \text{g}}{\cancel{\text{cm}}^3} = 780\ \text{g}$

(e) $275\ \cancel{\text{L}} \times \dfrac{1000\ \cancel{\text{mL}}}{1\ \cancel{\text{L}}} \times \dfrac{1\ \cancel{\text{cm}}^3}{1\ \cancel{\text{mL}}} \times \dfrac{1.025\ \text{g}}{\cancel{\text{cm}}^3} = 2.82 \times 10^5\ \text{g}$

37. (a) $\dfrac{19.7\ \text{g}}{25.0\ \text{mL}} = 0.788\ \text{g/mL}$

(b) $\dfrac{10.0\ \text{g}}{14.0\ \text{mL}} = 0.714\ \text{g/mL}$

(c) $\frac{11.6 \text{ g}}{4.1 \text{ mL}} = 2.8 \text{ g/mL}$

(d) $\frac{131.5 \text{ g}}{(3.55 \cancel{\text{cm}})(2.50 \cancel{\text{cm}})(1.75 \cancel{\text{cm}})} \times \frac{1 \cancel{\text{cm}^3}}{1 \text{ mL}} = 8.47 \text{ g/mL}$

(e) $4.03 \cancel{\mu\text{m}} \times \frac{1 \text{ cm}}{1 \times 10^4 \cancel{\mu\text{m}}} = 4.03 \times 10^{-4} \text{ cm}$

$\frac{0.175 \text{ g}}{(5.00 \cancel{\text{cm}})(4.50 \cancel{\text{cm}})(4.03 \times 10^{-4} \cancel{\text{cm}})} \times \frac{1 \cancel{\text{cm}^3}}{1 \text{ mL}} = 19.3 \text{ g/mL}$

**Section 2.9** *Temperature*

39.

| | Scale | Freezing Point of Water |
|---|---|---|
| (a) | Fahrenheit | 32°F |
| (b) | Kelvin | 273 K |
| (c) | Celsius | 0°C |

41. (a) $[(100 - 32)\cancel{°\text{F}}]\frac{100°\text{C}}{180\cancel{°\text{F}}} = 38°\text{C}$

(b) $[(-215 - 32)\cancel{°\text{F}}]\frac{100°\text{C}}{180\cancel{°\text{F}}} = -137°\text{C}$

(c) $[(2165 - 32)\cancel{°\text{F}}]\frac{100°\text{C}}{180\cancel{°\text{F}}} = 1185°\text{C}$

(d) $[(-40 - 32)\cancel{°\text{F}}]\frac{100°\text{C}}{180\cancel{°\text{F}}} = -40°\text{C}$

43. (a) $\left(42\cancel{°\text{C}} \times \frac{1 \text{ K}}{\cancel{°\text{C}}}\right) + 273 \text{ K} = 315 \text{ K}$

(b) $\left(-20\cancel{°\text{C}} \times \frac{1 \text{ K}}{\cancel{°\text{C}}}\right) + 273 \text{ K} = 253 \text{ K}$

(c) $\left(495\cancel{°\text{C}} \times \frac{1 \text{ K}}{\cancel{°\text{C}}}\right) + 273 \text{ K} = 768 \text{ K}$

(d) $\left(-185\cancel{°\text{C}} \times \frac{1 \text{ K}}{\cancel{°\text{C}}}\right) + 273 \text{ K} = 88 \text{ K}$

**Section 2.10** *Heat and Specific Heat*

45. Temperature is a measure of the average kinetic energy in a system and heat is a measure of the total energy in a system. (For example, a large chemistry lecture room at 20°C has more heat energy than a small laboratory room at 20°C, even though the temperature is the same.)

47. $250.0 \cancel{\text{g}} \times \frac{1.000 \text{ cal}}{1 \cancel{\text{g}} \cdot \cancel{°\text{C}}} \times (100 - 23)\cancel{°\text{C}} = 19{,}000 \text{ cal} \; (1.9 \times 10^4 \text{ cal})$

49. $25.0 \cancel{\text{g}} \times \frac{0.108 \text{ cal}}{1 \cancel{\text{g}} \cdot \cancel{°\text{C}}} \times (50.0 - 25.0)\cancel{°\text{C}} = 67.5 \text{ cal}$

51. $\dfrac{25.1 \text{ cal}}{30.0 \text{ g } (54.9 - 27.7)°\text{C}} = 0.0308 \text{ cal/g} \cdot °\text{C}$

53. $\dfrac{75.6 \cancel{\text{cal}}}{(31.4 - 20.7)\cancel{°\text{C}}} \times \dfrac{1 \text{ g} \cdot \cancel{°\text{C}}}{0.125 \cancel{\text{cal}}} = 56.5 \text{ g}$

55. $\dfrac{35.2 \cancel{\text{cal}}}{10.5 \cancel{\text{g}}} \times \dfrac{1 \cancel{\text{g}} \cdot °\text{C}}{0.0566 \cancel{\text{cal}}} = 59.2 \text{ °C}$

**Section 2.11** *The International System of Measurement, SI*

57. (a) $21{,}500{,}000{,}000 \cancel{\text{g}} \times \dfrac{1 \text{ Eg}}{1 \times 10^{18} \cancel{\text{g}}} = 2.15 \times 10^{-8} \text{ Eg}$

(b) $0.000\,000\,000\,02 \cancel{\text{TL}} \times \dfrac{1 \times 10^{12} \text{ L}}{\cancel{\text{TL}}} = 20 \text{ L}$

(c) $9.83 \times 10^{-17} \cancel{\text{m}} \times \dfrac{1 \times 10^{18} \text{ am}}{\cancel{\text{m}}} = 98.3 \text{ am}$

(d) $1.04 \times 10^{14} \cancel{\text{fs}} \times \dfrac{1 \text{ s}}{1 \times 10^{15} \cancel{\text{fs}}} = 0.104 \text{ s}$

59. $\dfrac{13.6 \cancel{\text{g}}}{\cancel{1 \text{ mL}}} \times \dfrac{1 \text{ kg}}{1000 \cancel{\text{g}}} \times \dfrac{1 \cancel{\text{mL}}}{1 \cancel{\text{cm}^3}} \times \dfrac{100 \cancel{\text{cm}}}{1 \text{ m}} \times \dfrac{100 \cancel{\text{cm}}}{1 \text{ m}} \times \dfrac{100 \cancel{\text{cm}}}{1 \text{ m}}$
$= 13{,}600 \text{ kg/m}^3$

**General Exercises**

61. $300 \cancel{\text{cubits}} \times \dfrac{18 \cancel{\text{in.}}}{1 \cancel{\text{cubit}}} \times \dfrac{1 \text{ ft}}{12 \cancel{\text{in.}}} = 450 \text{ ft}$

$50 \cancel{\text{cubits}} \times \dfrac{18 \cancel{\text{in.}}}{1 \cancel{\text{cubit}}} \times \dfrac{1 \text{ ft}}{12 \cancel{\text{in.}}} = 75 \text{ ft}$

$30 \cancel{\text{cubits}} \times \dfrac{18 \cancel{\text{in.}}}{1 \cancel{\text{cubit}}} \times \dfrac{1 \text{ ft}}{12 \cancel{\text{in.}}} = 45 \text{ ft}$

Dimensions: 450 ft by 75 ft by 45 ft

63. (a) an infinite number (*exact by definition*) (b) 3 significant digits
(c) an infinite number (*exact by definition*) (d) 3 significant digits

65. $\dfrac{975 \cancel{\text{miles}}}{\cancel{1 \text{ hour}}} \times \dfrac{1 \cancel{\text{hour}}}{60 \cancel{\text{minutes}}} \times \dfrac{1 \cancel{\text{minute}}}{60 \text{ s}} \times \dfrac{1.61 \cancel{\text{km}}}{1 \cancel{\text{mile}}} \times \dfrac{1000 \text{ m}}{\cancel{\text{km}}}$
$= 436 \text{ m/s}$

67. $\dfrac{3.00 \times 10^8 \cancel{\text{m}}}{\cancel{1 \text{ s}}} \times \dfrac{1 \cancel{\text{s}}}{1380 \cancel{\text{kilocycles}}} \times \dfrac{100 \text{ cm}}{\cancel{\text{m}}} \times \dfrac{1 \cancel{\text{kilocycle}}}{1000 \text{ cycles}}$
$= 2.17 \times 10^4 \text{ cm/cycle}$

69. $40 \cancel{\text{megabytes}} \times \dfrac{1000 \cancel{\text{kilobytes}}}{\cancel{\text{megabyte}}} \times \dfrac{1 \text{ floppy disk}}{800 \cancel{\text{kilobytes}}} = 50 \text{ floppy disks}$

71. $\frac{1.00\ \cancel{g}}{\cancel{1\ mL}} \times \frac{1\ lb}{454\ \cancel{g}} \times \frac{1\ \cancel{mL}}{1\ \cancel{cm^3}} \times \left(\frac{2.54\ \cancel{cm}}{1\ \cancel{in.}}\right)^3 \times \left(\frac{12\ \cancel{in.}}{ft}\right)^3 = 62.4\ lb/ft^3$

73. $25.0\ \cancel{mL} \times 0.791 \times \frac{1.00\ g}{\cancel{mL}} = 19.8\ g$

75. To convert °C to °N: $(\_\_\_°C)\frac{10°N}{1°C} - 500°N = \text{Temperature in } °N$

(a) $(37\ \cancel{°C})\frac{10°N}{1\ \cancel{°C}} - 500°N = -130°N$

(b) $(-273\ \cancel{°C})\frac{10°N}{1\ \cancel{°C}} - 500°N = -3230°N$

77. (a) $500.0\ \cancel{g} \times \frac{1.000\ cal}{1\ \cancel{g}\cdot\cancel{°C}} \times (71 - 22)\cancel{°C} = 2.5 \times 10^4\ cal$

(b) $2.5 \times 10^4\ \cancel{cal} \times \frac{4.184\ J}{\cancel{cal}} = 1.0 \times 10^5\ J$

79. $15.5\ g \times \frac{0.125\ cal}{1\ g\cdot°C} \times (T_{final} - 28.9)°C = -32.9\ cal$

$T_{final} = \left(\frac{-32.9\ \cancel{cal}}{15.5\ \cancel{g}} \times \frac{1\ \cancel{g}\cdot°C}{0.125\ \cancel{cal}}\right) + 28.9°C = 11.9°C$

81. $3.26\ \cancel{years} \times \frac{365\ \cancel{days}}{\cancel{year}} \times \frac{24\ \cancel{hours}}{\cancel{day}} \times \frac{60\ \cancel{minutes}}{\cancel{hour}} \times \frac{60\ \cancel{s}}{\cancel{minute}} \times \frac{3.00 \times 10^8\ m}{\cancel{s}} = 3.08 \times 10^{16}\ m$

# Matter and Energy

CHAPTER 3

## Section 3.1 *Physical States of Matter*

1.

| | State | Shape | Volume |
|---|---|---|---|
| (a) | solids | definite | fixed |
| (b) | liquids | indefinite | fixed |
| (c) | gases | indefinite | variable |

3.

| | State | Movement of Particles |
|---|---|---|
| (a) | solids | negligible |
| (b) | liquids | restricted |
| (c) | gases | unrestricted |

5.

| | Change in State | Term |
|---|---|---|
| (a) | liquid to gas | vaporizing |
| (b) | solid to gas | sublimation |
| (c) | liquid to solid | freezing |
| (d) | solid to liquid | melting |
| (e) | gas to liquid | condensing |
| (f) | gas to solid | deposition |

## Section 3.2 *Physical State and Energy*

7. Temperature is a measure of the average kinetic energy of a substance.

9. (a) The kinetic energy increases because temperature and kinetic energy are directly related.
   (b) The velocity increases because the kinetic energy is greater.

11. At the same temperature, the kinetic energy of lighter gas molecules is *equal to* the kinetic energy of heavier gas molecules.

13. If two gases have the same kinetic energy, the average temperature of lighter gas molecules is *equal to* the average temperature of heavier gas molecules.

## Section 3.3 *Classification of Matter*

15. (a) Homogeneous substances have definite properties and constant composition while heterogeneous mixtures have indefinite properties and variable composition.

    homogeneous substance: iron

    heterogeneous mixture: iron ore

    (b) Homogeneous mixtures have definite properties and variable composition while heterogeneous mixtures have indefinite properties and variable composition.

    homogeneous mixture: a penny (an alloy of copper and zinc)

    heterogeneous mixture: a container of copper and zinc shavings

    (c) An element is a pure substance that cannot be broken down any further by chemical reaction while a compound is a pure substance that can be broken down into elements by chemical reaction.

    element: sodium

    compound: table salt (sodium chloride)

17.

| | Substance | Classification | | Substance | Classification |
|---|---|---|---|---|---|
| (a) | ice | compound | (b) | distilled water | compound |
| (c) | silicon | element | (d) | Lake Mich. water | mixture |
| (e) | 18K jewelry | mixture | (f) | aluminum | element |
| (g) | stainless steel | mixture | (h) | uranium ore | mixture |

## Section 3.4 *Names and Symbols of the Elements*

19. The three most abundant elements in the earth's crust are oxygen, silicon, and aluminum.

21.

| | Element | Symbol | | Element | Symbol |
|---|---|---|---|---|---|
| (a) | bromine | Br | (b) | oxygen | O |
| (c) | antimony | Sb | (d) | lithium | Li |
| (e) | argon | Ar | (f) | magnesium | Mg |
| (g) | fluorine | F | (h) | sodium | Na |
| (i) | bismuth | Bi | (j) | nickel | Ni |
| (k) | helium | He | (l) | tellurium | Te |
| (m) | hydrogen | H | (n) | tin | Sn |
| (o) | calcium | Ca | (p) | platinum | Pt |
| (q) | carbon | C | (r) | potassium | K |
| (s) | iodine | I | (t) | titanium | Ti |
| (u) | iron | Fe | (v) | xenon | Xe |
| (w) | krypton | Kr | (x) | zinc | Zn |

23.

| | Element | Atomic # | | Element | Atomic # |
|---|---|---|---|---|---|
| (a) | bismuth | 83 | (b) | silver | 47 |
| (c) | cadmium | 48 | (d) | sodium | 11 |
| (e) | mercury | 80 | (f) | copper | 29 |
| (g) | potassium | 19 | (h) | tin | 50 |

**Section 3.5** *Metals, Nonmetals, and Semimetals*

25. The six semimetals that are not radioactive are: B, Si, Ge, As, Sb, Te

27.

| | Element | Physical State | | Element | Physical State |
|---|---|---|---|---|---|
| (a) | Li | solid | (b) | N | gas |
| (c) | P | solid | (d) | Sn | solid |
| (e) | Cd | solid | (f) | Mn | solid |
| (g) | Mg | solid | (h) | K | solid |
| (i) | Hg | liquid | (j) | Br | liquid |

29.

| | Element | Classification | | Element | Classification |
|---|---|---|---|---|---|
| (a) | aluminum | metal | (b) | boron | semimetal |
| (c) | phosphorus | nonmetal | (d) | manganese | metal |
| (e) | beryllium | metal | (f) | krypton | nonmetal |
| (g) | radium | metal | (h) | fluorine | nonmetal |
| (i) | hydrogen | nonmetal | (j) | uranium | metal |

**Section 3.6** *Compounds and Chemical Formulas*

31. The mass ratio of the elements in copper carbonate is 5 : 1 : 4 by mass. (From the law of constant composition)

33.*

| | Chemical Formula | # of Constituent Atoms |
|---|---|---|
| (a) | $C_9H_8O_4$ | 9 carbon atoms<br>8 hydrogen atoms<br>4 oxygen atoms |
| (b) | $C_{12}H_{22}O_{11}$ | 12 carbon atoms<br>22 hydrogen atoms<br>11 oxygen atoms |
| (c) | $C_3H_5(OH)_3$ | 3 carbon atoms<br>8 hydrogen atoms<br>3 oxygen atoms |

* continued on next page

33.

| | Chemical Formula | # of Constituent Atoms |
|---|---|---|
| (d) | $C_6H_2(NO_3)_3OH$ | 6 carbon atoms<br>3 hydrogen atoms<br>10 oxygen atoms<br>3 nitrogen atoms |
| (e) | $(C_2H_5)_2O$ | 4 carbon atoms<br>10 hydrogen atoms<br>1 oxygen atom |

35.

| | Chemical Formula | Total # of Atoms |
|---|---|---|
| (a) | $C_{27}H_{44}O$ | 72 |
| (b) | $C_8H_{17}N$ | 26 |
| (c) | $C_{20}H_{24}N_2O_2$ | 48 |
| (d) | $C_2H_4(OH)_2$ | 10 |
| (e) | $HO_2C(CH_2)_4CO_2H$ | 20 |

**Section 3.7** *Physical and Chemical Properties*

37.

| | Property | | Property |
|---|---|---|---|
| (a) | physical | (b) | chemical |
| (c) | physical | (d) | chemical |
| (e) | physical | | |

39.

| | Property | | Property |
|---|---|---|---|
| (a) | chemical | (b) | physical |
| (c) | chemical | (d) | physical |
| (e) | chemical | (f) | physical |
| (g) | physical | (h) | physical |
| (i) | physical | (j) | chemical |

**Section 3.8** *Physical and Chemical Changes*

41.

| | Change | | Change |
|---|---|---|---|
| (a) | physical | (b) | physical |
| (c) | physical | (d) | physical |
| (e) | chemical | (f) | physical |
| (g) | physical | | |

43.

| | Change | | Change |
|---|---|---|---|
| (a) | chemical | (b) | chemical |
| (c) | physical | (d) | chemical |
| (e) | physical | (f) | physical |
| (g) | physical | (h) | chemical |
| (i) | chemical | (j) | physical |

**Section 3.9** *Law of Conservation of Mass*

45. 2.50 g iron + 1.44 g sulfur = 3.94 g of iron sulfide

47. 0.750 g mercury oxide – 0.695 g residue = 0.055 g oxygen

**Section 3.10** *Law of Conservation of Energy*

49. On a roller coaster ride the potential energy is greatest when the car is at its highest point. As the car descends it loses potential energy and gains kinetic energy, having its greatest kinetic energy at its lowest point. As the car ascends it loses kinetic energy and gains potential energy.

51.

| | Type of Energy | | Type of Energy |
|---|---|---|---|
| (a) | potential | (b) | kinetic |
| (c) | potential | (d) | kinetic |
| (e) | potential | (f) | kinetic |
| (g) | kinetic | (h) | potential |
| (i) | potential | (j) | kinetic |

53. 697 calories + 110 calories = 807 calories released

55. Forms of Energy

(1) heat
(2) light
(3) chemical
(4) electrical
(5) mechanical
(6) nuclear

**Section 3.11** *Law of Conservation of Mass and Energy*

57. $E = mc^2$; E = energy, m = mass, c = the speed of light ($3.00 \times 10^8$ m/s)

59. The total amount of mass and energy for all substances before a chemical reaction is exactly equal to the total of mass and energy following the chemical reaction.

**General Exercises**

61. At high temperatures plasma is formed when electrons are stripped from atoms. The surface of the sun at 6000°C sends out solar flares of plasma.

63. $KE = \frac{1}{2}mv^2$; KE = kinetic energy of particle, m = mass of particle, v = velocity of particle

65. (a) homogeneous mixture (b) heterogeneous mixture
(c) homogeneous mixture (d) homogeneous mixture
(e) heterogeneous mixture

67. $1\ \cancel{\text{g BN}} \times \dfrac{3.5\ \text{kcal}}{1\ \cancel{\text{g BN}}} = 3.5\ \text{kcal}$

From the conservation of energy law, we know that the heat necessary to produce one gram of BN must be equal to the energy released when one gram of BN decomposes (3.5 kcal).

69.

| | Element | English Name | Symbol |
|---|---|---|---|
| (a) | hydrargyrum | mercury | Hg |
| (b) | natrium | sodium | Na |
| (c) | cuprum | copper | Cu |
| (d) | kalium | potassium | K |
| (e) | stibium | antimony | Sb |
| (f) | plumbum | lead | Pb |
| (g) | ferrum | iron | Fe |
| (h) | aurum | gold | Au |
| (i) | stannum | tin | Sn |
| (j) | argentum | silver | Ag |

# Atomic Theory and Structure

CHAPTER 4

**Section 4.1** *Evidence for Atoms: The Dalton Model*

1. (1) An element is composed of tiny, indivisible, indestructible particles called atoms.
   (2) All atoms of an element are identical and have the same properties.
   (3) Atoms of different elements combine to form compounds.
   (4) Compounds contain atoms in small, whole-number ratios.
   (5) Atoms may combine in more than one ratio to form different compounds.

3. (a) Atoms are divisible. (subatomic particles have been discovered)
   (b) All atoms of an element are not identical because elements may have isotopes.

**Section 4.2** *Evidence for Subatomic Particles: The Thomson Model*

5. The simplest particle observed in cathode rays was the electron ($e^-$).

7.

| Relative Charge | Relative Mass |
|---|---|
| 1– | 1/1836 the mass of a proton |

9. Raisins in the plum pudding model were analogous to electrons in atoms.

**Section 4.3** *Evidence for a Nuclear Atom: The Rutherford Model*

11. Rutherford concluded that atoms could not be 'plum-pudding' spheres as Thomson had previously suggested. Rutherford proposed that the mass of an atom exists only in a tiny fraction of its total volume. Furthermore, the mass is located at the center of the atom (nucleus) and has a positive charge.

13. In the planetary model of the atom, protons and neutrons are found at the center of the atom, known as the nucleus, while the electrons revolve in orbits about the nucleus.

15.

| Particle | Relative Charge |
|---|---|
| electron ($e^-$) | 1– |
| proton ($p^+$) | 1+ |
| neutron ($n^\circ$) | 0 |

**Section 4.4** *Atomic Notation*

17.

| | Isotope | Neutrons | | Isotope | Neutrons |
|---|---|---|---|---|---|
| (a) | $^{4}He$ | $4 - 2 = 2\ n^\circ$ | (b) | $^{32}S$ | $32 - 16 = 16\ n^\circ$ |
| (c) | $^{10}B$ | $10 - 5 = 5\ n^\circ$ | (d) | $^{44}Ca$ | $44 - 20 = 24\ n^\circ$ |
| (e) | $^{15}N$ | $15 - 7 = 8\ n^\circ$ | (f) | $^{52}Cr$ | $52 - 24 = 28\ n^\circ$ |
| (g) | $^{26}Mg$ | $26 - 12 = 14\ n^\circ$ | (h) | $^{58}Ni$ | $58 - 28 = 30\ n^\circ$ |

19.

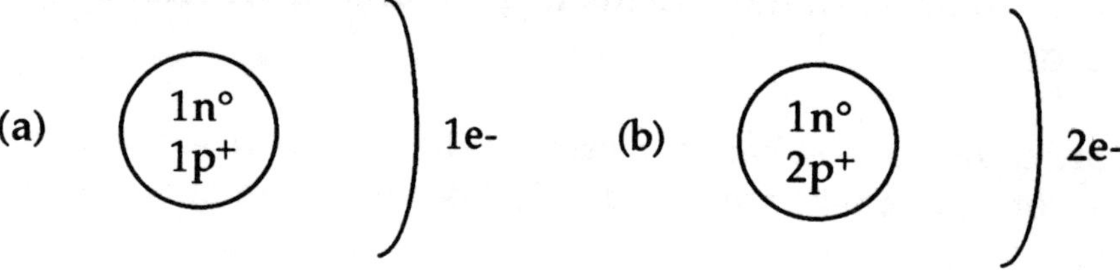


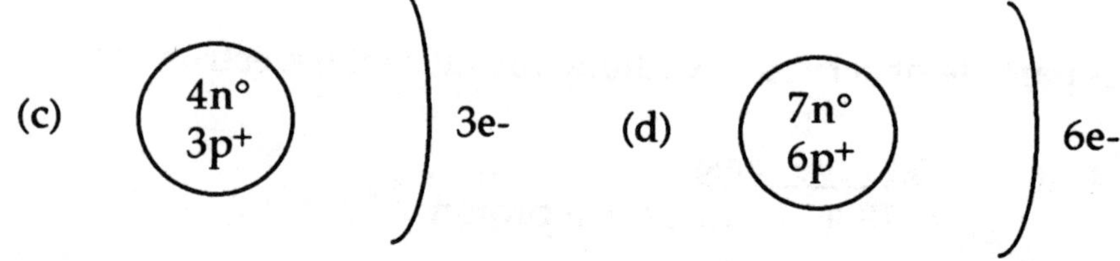


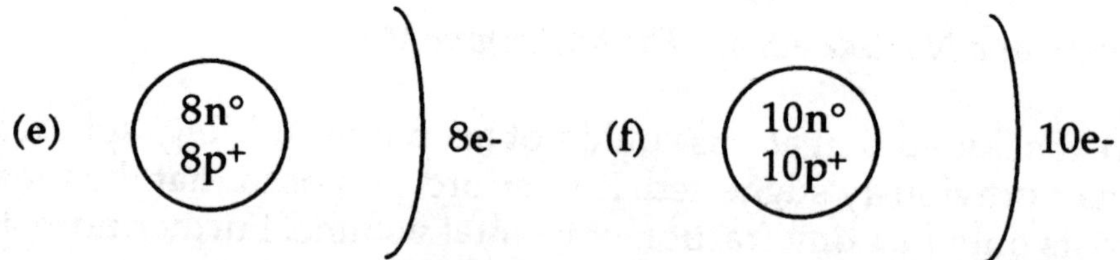


21. No, atomic numbers depend upon the number of protons in the nucleus of an atom and each element has a characteristic number of protons.

23.

| Atomic Notation | Atomic Number | Mass Number | Number of Protons | Number of Neutrons | Number of Electrons |
|---|---|---|---|---|---|
| $^{4}_{2}He$ | 2 | 4 | 2 | 2 | 2 |
| $^{13}_{6}C$ | 6 | 13 | 6 | 7 | 6 |
| $^{21}_{10}Ne$ | 10 | 21 | 10 | 11 | 10 |
| $^{28}_{14}Si$ | 14 | 28 | 14 | 14 | 14 |
| $^{40}_{18}Ar$ | 18 | 40 | 18 | 22 | 18 |
| $^{50}_{22}Ti$ | 22 | 50 | 22 | 28 | 22 |

**Section 4.5** *Atomic Mass Scale*

25. | Reference Isotope | Mass |
|---|---|
| carbon-12 | 12 amu (exactly) |

27.

| | Isotope | Magnetic Field | | Isotope | Magnetic Field |
|---|---|---|---|---|---|
| (a) | $^{4}_{2}He$ | decreased | (b) | $^{20}_{10}Ne$ | increased |
| (c) | $^{11}_{5}B$ | decreased | (d) | $^{14}_{6}C$ | increased |

**Section 4.6** *Atomic Mass*

29. The atomic mass of lithium, 6.941 amu, is the *weighted* average of the naturally occurring isotopic masses ($^{6}Li$ = 6.015 amu, $^{7}Li$ = 7.016 amu).

31. Simple Average Mass: $\frac{5\text{ g} + 2\text{ g}}{2} = 3.5\text{ g}$

Weighted Average Mass: $5\text{ g}\left(\frac{100\text{ lg. marbles}}{300\text{ total marbles}}\right) + 2\text{ g}\left(\frac{200\text{ sm. marbles}}{300\text{ total marbles}}\right) = 3\text{ g}$

33. Ag-107: 106.905 amu × 0.5182 = 55.40 amu

Ag-109: 108.905 amu × 0.4818 = 52.47 amu

*Atomic Mass* = 107.87 amu

| 35. | Ti-46: | 45.953 amu × 0.0793 | = | 3.64 amu |
|---|---|---|---|---|
| | Ti-47: | 46.952 amu × 0.0728 | = | 3.42 amu |
| | Ti-48: | 47.950 amu × 0.7394 | = | 35.45 amu |
| | Ti-49: | 48.948 amu × 0.0551 | = | 2.70 amu |
| | Ti-50: | 49.945 amu × 0.0534 | = | 2.67 amu |
| | | *Atomic Mass* | = | 47.88 amu |

| 37. | Mg-24: | 23.985 amu × 0.7870 | = | 18.88 amu |
|---|---|---|---|---|
| | Mg-25: | 24.986 amu × 0.1013 | = | 2.531 amu |
| | Mg-26: | 25.983 amu × 0.1117 | = | 2.902 amu |
| | | *Atomic Mass* | = | 24.31 amu |

39. Cl-35 must be more stable because the atomic weight 35.5 amu indicates that Cl-35 has a greater percent abundance than Cl-37 does.

41. Polonium is radioactive because the periodic table lists a mass number (209) rather than an atomic mass. (All elements that list whole-numbers below the symbol are radioactive.)

**Section 4.7** *Evidence for Electron Energy Levels: The Bohr Model*

43. Violet light is more energetic than green or orange light.

45. Light having a wavelength of 450 nm is more energetic than light with a wavelength of either 550 nm or 650 nm. (Lower wavelengths of light correspond to greater energy.)

47. The Bohr Planetary Model of the Atom

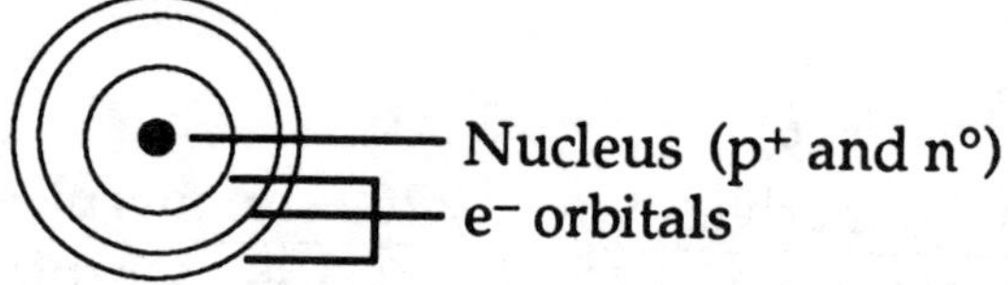


49. Excited electrons that return to the ground state release heat and light energy.

51.

| | Color | Energy Level |
|---|---|---|
| (a) | red | 3 to 2 |
| (b) | blue-green | 4 to 2 |
| (c) | violet | 5 to 2 |

53. The 5 to 2 transition is the most energetic because the electron drops the greatest distance.

55. 4 to 2; $n = 4$

$$\frac{1}{\lambda} = \frac{1}{91\text{nm}}\left(\frac{1}{2^2} - \frac{1}{n^2}\right)$$

$$\frac{1}{\lambda} = \frac{1}{91\text{nm}}\left(\frac{1}{4} - \frac{1}{16}\right)$$

$$\frac{1}{\lambda} = \frac{1}{91\text{nm}}(0.25 - 0.0625)$$

$$\frac{1}{\lambda} = \frac{0.1875}{91\text{ nm}}$$

$$\lambda = \frac{91\text{ nm}}{0.1875} = 490\text{ nm (Only two significant digits)}$$

57. 5 to 2; $n = 5$

$$\frac{1}{\lambda} = \frac{1}{91\text{nm}}\left(\frac{1}{2^2} - \frac{1}{n^2}\right)$$

$$\frac{1}{\lambda} = \frac{1}{91\text{nm}}\left(\frac{1}{4} - \frac{1}{25}\right)$$

$$\frac{1}{\lambda} = \frac{1}{91\text{nm}}(0.25 - 0.04)$$

$$\frac{1}{\lambda} = \frac{0.21}{91\text{ nm}}$$

$$\lambda = \frac{91\text{ nm}}{0.21} = 430\text{ nm (Only two significant digits)}$$

**Section 4.8** *Principal Energy Levels and Sublevels*

59. The lines in the emission spectrum of hydrogen suggested the existence of energy levels.

61.

| | Principal Energy Level | Sublevels |
|---|---|---|
| (a) | first | 1*s* |
| (b) | second | 2*s* 2*p* |
| (c) | third | 3*s* 3*p* 3*d* |
| (d) | fourth | 4*s* 4*p* 4*d* 4*f* |

63.

| | Sublevel | Max. # of Electrons | | Sublevel | Max. # of Electrons |
|---|---|---|---|---|---|
| (a) | 2*s* | 2 $e^-$ | (b) | 3*s* | 2 $e^-$ |
| (c) | 2*p* | 6 $e^-$ | (d) | 4*p* | 6 $e^-$ |
| (e) | 3*d* | 10 $e^-$ | (f) | 5*d* | 10 $e^-$ |
| (g) | 4*f* | 14 $e^-$ | (h) | 5*f* | 14 $e^-$ |

65. 

| | Principal Energy Level | Max. # of Electrons |
|---|---|---|
| (a) | first | 2 e⁻ |
| (b) | second | 2 e⁻ + 6 e⁻ = 8 e⁻ |
| (c) | third | 2 e⁻ + 6 e⁻ + 10 e⁻ = 18 e⁻ |
| (d) | fourth | 2 e⁻ + 6 e⁻ + 10 e⁻ + 14 e⁻ = 32 e⁻ |

**Section 4.9** *Arrangement of Electrons by Energy Sublevel*

67. Order: 1*s* 2*s* 2*p* 3*s* 3*p* 4*s* 3*d* 4*p* 5*s* 4*d* 5*p*

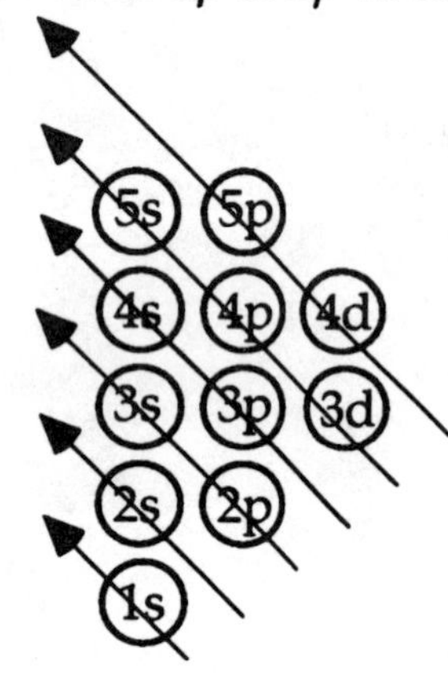


69.

| | Element | Electron Configuration |
|---|---|---|
| (a) | He | $1s^2$ |
| (b) | Be | $1s^2\ 2s^2$ |
| (c) | C | $1s^2\ 2s^2\ 2p^2$ |
| (d) | Mg | $1s^2\ 2s^2\ 2p^6\ 3s^2$ |
| (e) | S | $1s^2\ 2s^2\ 2p^6\ 3s^2\ 3p^4$ |
| (f) | K | $1s^2\ 2s^2\ 2p^6\ 3s^2\ 3p^6\ 4s^1$ |
| (g) | Co | $1s^2\ 2s^2\ 2p^6\ 3s^2\ 3p^6\ 4s^2\ 3d^7$ |
| (h) | Cd | $1s^2\ 2s^2\ 2p^6\ 3s^2\ 3p^6\ 4s^2\ 3d^{10}\ 4p^6\ 5s^2\ 4d^{10}$ |

71.

| | Electron Configuration | Element |
|---|---|---|
| (a) | $1s^2\ 2s^1$ | Li |
| (b) | $1s^2\ 2s^2\ 2p^6\ 3s^2\ 3p^2$ | Si |
| (c) | $1s^2\ 2s^2\ 2p^6\ 3s^2\ 3p^6\ 4s^2\ 3d^2$ | Ti |
| (d) | $1s^2\ 2s^2\ 2p^6\ 3s^2\ 3p^6\ 4s^2\ 3d^{10}\ 4p^6\ 5s^2$ | Sr |

## General Exercises

73. If a missile represents an alpha particle, then the planets represent atomic nuclei.

75. $1.60 \times 10^{-19}\, \cancel{\text{coulomb}} \times \dfrac{1\text{ g}}{1.76 \times 10^{8}\, \cancel{\text{coulomb}}} = 9.09 \times 10^{-28}\text{ g}$

77. Diameter of atom: $\sim 10^{-8}$ cm
Diameter of golf ball: $\sim 4$ cm
Size of golf ball vs. atom: $\dfrac{4\, \cancel{\text{cm}}}{10^{-8}\, \cancel{\text{cm}}} = 4 \times 10^{8}$ times larger
Magnification of golf ball: $4 \times 10^{8} \times 4\, \cancel{\text{cm}} \times \dfrac{1\, \cancel{\text{m}}}{100\, \cancel{\text{cm}}} \times \dfrac{1\text{ km}}{1000\, \cancel{\text{m}}}$
$\approx 12{,}000$ km

Thus, if a golf ball was magnified the same number of times as enlarging an atom to the size of a golf ball, then the golf ball would be approximately the size of the earth.

79. An algebraic method must be used to solve this problem.

Let **X** = the unknown isotopic mass of Ga-71
Percent abundance of Ga-71: 100% – 60.1% = 39.9%

| | | | |
|---|---|---|---|
| Ga-69: | 68.92558 amu × 0.601 | = | 41.4 amu |
| Ga-71: | **X** × 0.399 | = | 0.399 **X** |
| | *Atomic Mass* | = | 69.723 amu |

41.4 amu + 0.399 **X** = 69.723 amu

Solve for **X**: 0.399 **X** = 69.723 amu – 41.4 amu
0.399 **X** = 28.3 amu
$\mathbf{X} = \dfrac{28.3\text{ amu}}{0.399} = 70.9\text{ amu}$

The isotopic mass of Ga-71 = 70.9 amu

81. No, neon always emits a reddish-orange light. Other gases, for example helium, argon, and krypton, emit different colors.

83.

| | Wavelength | Region | | Wavelength | Region |
|---|---|---|---|---|---|
| (a) | 200 nm | ultraviolet | (b) | 500 nm | visible |
| (c) | 1200 nm | infrared | (d) | 1500 nm | infrared |

85. (Note: The designation $n_L$ denotes the lower quantum level and $n_H$ denotes the higher quantum level.)

2 to 1; $n_L = 1$; $n_H = 2$

$$\frac{1}{\lambda} = \frac{1}{91\text{nm}}\left(\frac{1}{n_L{}^2} - \frac{1}{n_H{}^2}\right)$$

$$\frac{1}{\lambda} = \frac{1}{91\text{nm}}\left(\frac{1}{1^2} - \frac{1}{2^2}\right)$$

$$\frac{1}{\lambda} = \frac{1}{91\text{nm}}(1 - 0.25)$$

$$\frac{1}{\lambda} = \frac{0.75}{91\text{ nm}}$$

$$\lambda = \frac{91\text{ nm}}{0.75} = 120\text{ nm (Only two significant digits)}$$

87. 4 to 3; $n_L = 3$; $n_H = 4$

$$\frac{1}{\lambda} = \frac{1}{91\text{nm}}\left(\frac{1}{n_L{}^2} - \frac{1}{n_H{}^2}\right)$$

$$\frac{1}{\lambda} = \frac{1}{91\text{nm}}\left(\frac{1}{3^2} - \frac{1}{4^2}\right)$$

$$\frac{1}{\lambda} = \frac{1}{91\text{nm}}(0.1111 - 0.0625)$$

$$\frac{1}{\lambda} = \frac{0.0486}{91\text{ nm}}$$

$$\lambda = \frac{91\text{ nm}}{0.0486} = 1900\text{ nm (Only two significant digits)}$$

89. The third main energy level can hold a maximum of 18 electrons because of the 3*s*, 3*p*, and 3*d* sublevels. Yet, the ten 3*d* electrons are not found in the third row of the periodic table, they are found in the fourth row.

# The Periodic Table

CHAPTER 5

**Section 5.1** *Systematic Arrangement of the Elements*

1. Between chlorine and iodine: bromine (Br)
   Between calcium and barium: strontium (Sr)

3. Following the potassium series: copper (Cu)
   Following the rubidium series: silver (Ag)

**Section 5.2** *The Periodic Law*

5. Newlands, Meyer, and Mendeleev suggested that the elements should be arranged according to increasing atomic weight.

7. Based on the periodic law, the chemical and physical properties of the elements tend to recur periodically when the elements are arranged according to increasing atomic number.

**Section 5.3** *The Periodic Table of the Elements*

9. Vertical columns in the periodic table are referred to as chemical groups and chemical families.

11. The main group elements, Groups IA through VIIIA, are called the representative elements.

13. The two series of elements that include Ce-Lu and Th-Lr are known as the inner transition elements.

15. The elements on the right side of the periodic table are called the nonmetals.

17.

| | Family | Group Number |
|---|---|---|
| (a) | alkali metals | IA/1 |
| (b) | alkaline earth metals | IIA/2 |
| (c) | halogens | VIIA/17 |
| (d) | noble gases | VIIIA/18 |

19. The elements in the series that follow element 57 are called the lanthanide elements.

21. The Group IIIB/3 elements are the rare earth elements.

23.

| | American | IUPAC | | American | IUPAC |
|---|---|---|---|---|---|
| (a) | Group IA | 1 | (b) | Group IB | 11 |
| (c) | Group IIIA | 13 | (d) | Group IIIB | 3 |
| (e) | Group VA | 15 | (f) | Group VB | 5 |
| (g) | Group VIIA | 17 | (h) | Group VIIB | 7 |

25.

| | Element | | Element |
|---|---|---|---|
| (a) | Ge | (b) | Na |
| (c) | I | (d) | Pm |
| (e) | Lu | (f) | Ba |
| (g) | Zr | (h) | He |

**Section 5.4** *Periodic Trends*

27. Proceeding down a group of elements in the periodic table, the atomic radius of an element generally increases.

29. Proceeding down a group of elements in the periodic table, the metallic character of an element generally increases.

31.

| | Atomic Radius | | Atomic Radius |
|---|---|---|---|
| (a) | Li < **Na** | (b) | N < **P** |
| (c) | Mg < **Ca** | (d) | Ar < **Kr** |
| (e) | **Rb** > Sr | (f) | **As** > Se |
| (g) | **Pb** > Bi | (h) | **I** > Xe |

(Note: The element in bold indicates the larger atomic radius.)

**Section 5.5** *Predicting Properties of Elements*

33. Predicted atomic radius of K: $0.248 - (0.266 - 0.248) = \sim 0.230$ nm

Predicted density of Rb: $\frac{0.86 + 1.90}{2} = \sim 1.38$ g/mL

Predicted melting point of Cs: $38.9 - (63.3 - 38.9) = \sim 14.5$°C

35. Predicted atomic radius of Cr: $0.131 - (0.137 - 0.131) = \sim 0.125$ nm

Predicted density of Mo: $\dfrac{19.26 + 7.14}{2} = \sim 13.20$ g/mL

Predicted melting point of W: $2617 + (2617 - 1857) = \sim 3377$°C

37.

| | Compound | Formula | | Compound | Formula |
|---|---|---|---|---|---|
| (a) | lithium oxide | $Li_2O$ | (b) | calcium oxide | $CaO$ |
| (c) | gallium oxide | $Ga_2O_3$ | (d) | tin oxide | $SnO_2$ |

39.

| | Compound | Formula | | Compound | Formula |
|---|---|---|---|---|---|
| (a) | cadmium oxide | $CdO$ | (b) | zinc sulfide | $ZnS$ |
| (c) | mercury sulfide | $HgS$ | (d) | cadmium selenide | $CdSe$ |

41.

| | Compound | Formula | | Compound | Formula |
|---|---|---|---|---|---|
| (a) | sulfur oxide | $SO_3$ | (b) | tellurium oxide | $TeO_3$ |
| (c) | selenium sulfide | $SeS_3$ | (d) | tellurium sulfide | $TeS_3$ |

## Section 5.6 *The s, p, d, f, Blocks of Elements*

43. The *s* sublevel is filled by the elements in Groups IA/1 and IIA/2.

45. The *d* sublevel is filled by the elements in Groups IIIB/3 through IIB/12.

47. The 4*f* sublevel is filled by the lanthanide series.

49.

| | Element | Highest Sublevel | | Element | Highest Sublevel |
|---|---|---|---|---|---|
| (a) | H | 1*s* | (b) | Na | 3*s* |
| (c) | Sm | 4*f* | (d) | Br | 4*p* |
| (e) | Sr | 5*s* | (f) | C | 2*p* |
| (g) | Sn | 5*p* | (h) | Cs | 6*s* |
| (i) | Tl | 6*p* | (j) | Unq | 6*d* |

51.*

| | Element | Electron Configuration | Core Notation |
|---|---|---|---|
| (a) | Li | $1s^2\ 2s^1$ | $[He]\ 2s^1$ |
| (b) | F | $1s^2\ 2s^2\ 2p^5$ | $[He]\ 2s^2\ 2p^5$ |
| (c) | Mg | $1s^2\ 2s^2\ 2p^6\ 3s^2$ | $[Ne]\ 3s^2$ |
| (d) | P | $1s^2\ 2s^2\ 2p^6\ 3s^2\ 3p^3$ | $[Ne]\ 3s^2\ 3p^3$ |
| (e) | Ca | $1s^2\ 2s^2\ 2p^6\ 3s^2\ 3p^6\ 4s^2$ | $[Ar]\ 4s^2$ |
| (f) | Mn | $1s^2\ 2s^2\ 2p^6\ 3s^2\ 3p^6\ 4s^2\ 3d^5$ | $[Ar]\ 4s^2\ 3d^5$ |
| (g) | Ga | $1s^2\ 2s^2\ 2p^6\ 3s^2\ 3p^6\ 4s^2\ 3d^{10}\ 4p^1$ | $[Ar]\ 4s^2\ 3d^{10}\ 4p^1$ |

* continued on next page

51.

| | Element | Electron Configuration | Core Notation |
|---|---|---|---|
| (h) | Rb | $1s^2\,2s^2\,2p^6\,3s^2\,3p^6\,4s^2\,3d^{10}\,4p^6\,5s^1$ | [Kr] $5s^1$ |
| (i) | Tc | $1s^2\,2s^2\,2p^6\,3s^2\,3p^6\,4s^2\,3d^{10}\,4p^6\,5s^2\,4d^5$ | [Kr] $5s^2\,4d^5$ |
| (j) | Xe | $1s^2\,2s^2\,2p^6\,3s^2\,3p^6\,4s^2\,3d^{10}\,4p^6\,5s^2\,4d^{10}\,5p^6$ | [Kr] $5s^2\,4d^{10}\,5p^6$ |

**Section 5.7** *Predicting Valence Electrons*

53.

| | Group | Valence Electrons | | Group | Valence Electrons |
|---|---|---|---|---|---|
| (a) | IA/1 | 1 | (b) | IIIA/3 | 3 |
| (c) | VA/15 | 5 | (d) | VIIA/17 | 7 |

55.

| | Element | Valence Electrons | | Element | Valence Electrons |
|---|---|---|---|---|---|
| (a) | H | 1 | (b) | B | 3 |
| (c) | N | 5 | (d) | F | 7 |
| (e) | Ca | 2 | (f) | Si | 4 |
| (g) | O | 6 | (h) | Ar | 8 |

**Section 5.8** *Electron Dot Formulas of Atoms*

57.

| | Element | Electron Dot Formula | | Element | Electron Dot Formula |
|---|---|---|---|---|---|
| (a) | H | H· | (b) | B | ·Ḅ· |
| (c) | N | ·Ṇ̇: | (d) | F | :Ḟ̤: |
| (e) | Ca | Cạ· | (f) | Si | ·Ṩi· |
| (g) | O | ·Ȯ̤: | (h) | Ar | :Ä̤r: |

**Section 5.9** *Ionization Energy*

59. Proceeding down a group of elements in the periodic table, the first ionization energy generally decreases.

61. The Group VIIIA/18 elements have the highest ionization energy.

63.

| | Ionization Energy | | Ionization Energy |
|---|---|---|---|
| (a) | **Mg** > Ca | (b) | **S** > Se |
| (c) | **Sn** > Pb | (d) | **N** > P |
| (e) | Sn < **Sb** | (f) | Se < **Br** |
| (g) | Ga < **Ge** | (h) | Si < **P** |
| (i) | Br < **Cl** | (j) | **As** > Sb |

(Note: The element in bold has the higher ionization energy.)

**Section 5.10** *Predicting Ionic Charge*

65.

| | Group | Ionic Charge | | Group | Ionic Charge |
|---|---|---|---|---|---|
| (a) | IA/1 | 1+ | (b) | IIA/2 | 2+ |
| (c) | IIIA/13 | 3+ | (d) | IVA/14 | 4+ |

67.

| | Ion | Charge | | Ion | Charge |
|---|---|---|---|---|---|
| (a) | K ion | 1+ | (b) | Be ion | 2+ |
| (c) | Al ion | 3+ | (d) | Sn ion | 4+ |
| (e) | C ion | 4– | (f) | Se ion | 2– |
| (g) | O ion | 2– | (h) | F ion | 1– |

69.

| | Ion | Isoelectronic with: | | Ion | Isoelectronic with: |
|---|---|---|---|---|---|
| (a) | $K^+$ | Ar | (b) | $Ca^{2+}$ | Ar |
| (c) | $Al^{3+}$ | Ne | (d) | $Cl^-$ | Ar |
| (e) | $S^{2-}$ | Ar | (f) | $N^{3-}$ | Ne |

**Section 5.11** *Electron Configuration of Ions*

71.

| | Ion | Electron Configuration |
|---|---|---|
| (a) | $Mg^{2+}$ | $1s^2\,2s^2\,2p^6$ |
| (b) | $K^+$ | $1s^2\,2s^2\,2p^6\,3s^2\,3p^6$ |
| (c) | $Fe^{2+}$ | $1s^2\,2s^2\,2p^6\,3s^2\,3p^6\,3d^6$ |
| (d) | $Cs^+$ | $1s^2\,2s^2\,2p^6\,3s^2\,3p^6\,4s^2\,3d^{10}\,4p^6\,5s^2\,4d^{10}\,5p^6$ |

73.

| | Ion | Electron Configuration |
|---|---|---|
| (a) | $F^-$ | $1s^2\,2s^2\,2p^6$ |
| (b) | $S^{2-}$ | $1s^2\,2s^2\,2p^6\,3s^2\,3p^6$ |
| (c) | $N^{3-}$ | $1s^2\,2s^2\,2p^6$ |
| (d) | $I^-$ | $1s^2\,2s^2\,2p^6\,3s^2\,3p^6\,4s^2\,3d^{10}\,4p^6\,5s^2\,4d^{10}\,5p^6$ |

**General Exercises**

75. ekaboron: scandium
ekaaluminum: gallium
ekasilicon: germanium

77.

| | European | American | | European | American |
|---|---|---|---|---|---|
| (a) | Group IA | Group IA | (b) | Group IB | Group IB |
| (c) | Group IIIA | Group IIIB | (d) | Group IIIB | Group IIIA |
| (e) | Group VA | Group VB | (f) | Group VB | Group VA |

79. Predicted atomic radius of Fr: $0.266 + (0.266 - 0.248) = \sim 0.284$ nm

Predicted density of Fr: $1.87 + (1.87 - 1.53) = \sim 2.21$ g/mL

Predicted melting point of Fr: $28.4 - (38.9 - 28.4) = \sim 17.9°C$

81. Alkali metal atoms lose one electron and assume a noble gas configuration. Alkaline earth metals, after losing one electron, do not assume a noble gas electron configuration. However, the alkaline earth metal atoms become very stable after losing two electrons to attain a noble gas electron configuration.

83. Although both hydrogen and the other Group IA/1 elements each have one valence electron, the electron in hydrogen is in the 1*s* sublevel while the outer electrons of the other Group IA/1 elements are at higher *s* sublevels. As a result, the electron in hydrogen is closer to its nucleus than the valence electrons of the other elements. Because it is closer, the electron in hydrogen is more difficult to remove than the valence electrons in the other elements.

# Naming Chemical Compounds

CHAPTER 6

**Section 6.1** *IUPAC Systematic Nomenclature*

1.

| | Compound | Classification | | Compound | Classification |
|---|---|---|---|---|---|
| (a) | $AgNO_3$ | ternary ionic | (b) | $Li_3N$ | binary ionic |
| (c) | $NH_3$ | binary molecular | (d) | $HBr_{(aq)}$ | binary acid |
| (e) | $H_2SO_{3(aq)}$ | ternary oxyacid | | | |

3.

| | Ion | Classification | | Ion | Classification |
|---|---|---|---|---|---|
| (a) | $Br^-$ | monoatomic anion | (b) | $NH_4^+$ | polyatomic cation |
| (c) | $C_2H_3O_2^-$ | polyatomic anion | (d) | $Hg^{2+}$ | monoatomic cation |

**Section 6.2** *Monoatomic Ions*

5.

| | Monoatomic Cation | Systematic Name |
|---|---|---|
| (a) | $K^+$ | potassium ion |
| (b) | $Ba^{2+}$ | barium ion |
| (c) | $Ag^+$ | silver ion |
| (d) | $Cd^{2+}$ | cadmium ion |

7.

| | Monoatomic Cation | Stock System Name |
|---|---|---|
| (a) | $Hg^{2+}$ | mercury(II) ion |
| (b) | $Cu^{2+}$ | copper(II) ion |
| (c) | $Fe^{2+}$ | iron(II) ion |
| (d) | $Co^{3+}$ | cobalt(III) ion |

9.

| | Monoatomic Cation | Latin Name |
|---|---|---|
| (a) | $Cu^+$ | cuprous ion |
| (b) | $Fe^{3+}$ | ferric ion |
| (c) | $Sn^{2+}$ | stannous ion |
| (d) | $Pb^{4+}$ | plumbic ion |

11.

| | Monoatomic Anion | Systematic Name |
|---|---|---|
| (a) | $F^-$ | fluoride ion |
| (b) | $I^-$ | iodide ion |
| (c) | $O^{2-}$ | oxide ion |
| (d) | $P^{3-}$ | phosphide ion |

**Section 6.3** *Polyatomic Ions*

13.

| | Polyatomic Anion | Systematic Name |
|---|---|---|
| (a) | $ClO^-$ | hypochlorite ion |
| (b) | $SO_3^{2-}$ | sulfite ion |
| (c) | $C_2H_3O_2^-$ | acetate ion |
| (d) | $CO_3^{2-}$ | carbonate ion |

15.

| | Polyatomic Ion | Formula |
|---|---|---|
| (a) | hydroxide ion | $OH^-$ |
| (b) | nitrite ion | $NO_2^-$ |
| (c) | dichromate ion | $Cr_2O_7^{2-}$ |
| (d) | hydrogen carbonate ion | $HCO_3^-$ |

**Section 6.4** *Writing Chemical Formulas*

17.

| | Ions | | Binary Compound |
|---|---|---|---|
| (a) | $Na^+ + Cl^-$ | = | $NaCl$ |
| (b) | $Al^{3+} + 3\ Br^-$ | = | $AlBr_3$ |
| (c) | $2\ Ag^+ + O^{2-}$ | = | $Ag_2O$ |
| (d) | $2\ Bi^{3+} + 3\ O^{2-}$ | = | $Bi_2O_3$ |
| (e) | $Sn^{4+} + 4\ I^-$ | = | $SnI_4$ |

19.

| | Ions | | Ternary Compound |
|---|---|---|---|
| (a) | $K^+ + NO_3^-$ | = | $KNO_3$ |
| (b) | $2\ NH_4^+ + Cr_2O_7^{2-}$ | = | $(NH_4)_2Cr_2O_7$ |
| (c) | $2\ Al^{3+} + 3\ SO_3^{2-}$ | = | $Al_2(SO_3)_3$ |
| (d) | $Bi^{3+} + 3\ ClO^-$ | = | $Bi(ClO)_3$ |

21.

| | Ions | | Ternary Compound |
|---|---|---|---|
| (a) | $Sr^{2+} + 2\ NO_2^-$ | = | $Sr(NO_2)_2$ |
| (b) | $Zn^{2+} + 2\ MnO_4^-$ | = | $Zn(MnO_4)_2$ |
| (c) | $Ca^{2+} + CrO_4^{2-}$ | = | $CaCrO_4$ |
| (d) | $Cr^{3+} + 3\ ClO_4^-$ | = | $Cr(ClO_4)_3$ |

## Section 6.5 *Binary Ionic Compounds*

23.

| | Binary Ionic Compound | Systematic Name |
|---|---|---|
| (a) | $MgO$ | magnesium oxide |
| (b) | $AgBr$ | silver bromide |
| (c) | $CdCl_2$ | cadmium chloride |
| (d) | $Al_2S_3$ | aluminum sulfide |

25.

| | Binary Ionic Compound | Stock System Name |
|---|---|---|
| (a) | $CuO$ | copper(II) oxide |
| (b) | $FeO$ | iron(II) oxide |
| (c) | $HgO$ | mercury(II) oxide |
| (d) | $SnO$ | tin(II) oxide |

27.

| | Binary Ionic Compound | Latin Name |
|---|---|---|
| (a) | $Cu_2O$ | cuprous oxide |
| (b) | $Fe_2O_3$ | ferric oxide |
| (c) | $Hg_2O$ | mercurous oxide |
| (d) | $SnO_2$ | stannic oxide |

## Section 6.6 *Ternary Ionic Compounds*

29.

| | Ternary Ionic Compound | Systematic Name |
|---|---|---|
| (a) | $KMnO_4$ | potassium permanganate |
| (b) | $Sr(ClO_4)_2$ | strontium perchlorate |
| (c) | $BaCrO_4$ | barium chromate |
| (d) | $Cd(CN)_2$ | cadmium cyanide |

31.

| | Ternary Ionic Compound | Stock System Name |
|---|---|---|
| (a) | $CuSO_4$ | copper(II) sulfate |
| (b) | $FeCrO_4$ | iron(II) chromate |
| (c) | $Hg(NO_2)_2$ | mercury(II) nitrite |
| (d) | $Pb(C_2H_3O_2)_2$ | lead(II) acetate |

33.

| | Ternary Ionic Compound | Latin Name |
|---|---|---|
| (a) | $Cu_2SO_4$ | cuprous sulfate |
| (b) | $Fe_2(CrO_4)_3$ | ferric chromate |
| (c) | $Hg_2(NO_2)_2$ | mercurous nitrite |
| (d) | $Pb(C_2H_3O_2)_4$ | plumbic acetate |

**Section 6.7** *Binary Molecular Compounds*

35.

| | Binary Molecular Compound | Systematic Name |
|---|---|---|
| (a) | $S_2Cl_2$ | disulfur dichloride |
| (b) | $P_2O_3$ | diphosphorus trioxide |
| (c) | $N_2O$ | dinitrogen oxide |
| (d) | $C_3O_2$ | tricarbon dioxide |

37.

| | Binary Molecular Compound | Formula |
|---|---|---|
| (a) | nitrogen dioxide | $NO_2$ |
| (b) | carbon tetrachloride | $CCl_4$ |
| (c) | iodine monobromide | IBr |
| (d) | hydrogen sulfide | $H_2S$ |

**Section 6.8** *Binary Acids*

39.

| | Binary Acid | Systematic Name |
|---|---|---|
| (a) | $HF_{(aq)}$ | hydrofluoric acid |
| (b) | $HBr_{(aq)}$ | hydrobromic acid |
| (c) | $H_2Se_{(aq)}$ | hydroselenic acid |

**Section 6.9** *Ternary Oxyacids*

41.

| | Ternary Oxyacid | Systematic Name |
|---|---|---|
| (a) | $HClO_{2(aq)}$ | chlorous acid |
| (b) | $HNO_{3(aq)}$ | nitric acid |
| (c) | $H_3PO_{4(aq)}$ | phosphoric acid |
| (d) | $H_2SO_{3(aq)}$ | sulfurous acid |

43.

| | Ternary Oxyacid | Formula |
|---|---|---|
| (a) | acetic acid | $HC_2H_3O_{2(aq)}$ |
| (b) | hypochlorous acid | $HClO_{(aq)}$ |
| (c) | phosphorous acid | $H_3PO_{3(aq)}$ |
| (d) | chloric acid | $HClO_{3(aq)}$ |

**Section 6.10** *Acid Salts*

45.

| | Acids Salts | Systematic Name |
|---|---|---|
| (a) | $CaHPO_4$ | calcium hydrogen phosphate |
| (b) | $LiH_2PO_4$ | lithium dihydrogen phosphate |
| (c) | $Zn(HCO_3)_2$ | zinc hydrogen carbonate |
| (d) | $Ni(HSO_3)_2$ | nickel(II) hydrogen sulfite |

**Section 6.11** *Predicting Chemical Formulas*

47. (a) Given: sodium chloride = $NaCl$
rubidium chloride = $RbCl$
(b) Given: calcium oxide = $CaO$
beryllium oxide = $BeO$

49. (a) Given: sodium sulfate = $Na_2SO_4$
francium sulfate = $Fr_2SO_4$
(b) Given: barium chlorate = $Ba(ClO_3)_2$
barium bromate = $Ba(BrO_3)_2$

51. (a) Given: hypochlorous acid = $HClO_{(aq)}$
hypobromous acid = $HBrO_{(aq)}$
(b) Given: perchloric acid = $HClO_{4(aq)}$
periodic acid = $HIO_{4(aq)}$

**General Exercises**

53.

| | Substance | Ionic Charge |
|---|---|---|
| (a) | iron metal atoms | 0 |
| (b) | ferrous ions | $2+$ |
| (c) | iron (III) ions | $3+$ |
| (d) | iron compounds | 0 (the total ionic charge of the compound) |

55.

| | Polyatomic Anion | Valence Electrons |
|---|---|---|
| (a) | $IO_4^{?-}$ | $7 + 4(6) + (2\ e^-$ from charge$) = 33$ |
| (b) | $S_2O_3^{?-}$ | $2(6) + 3(6) + (2\ e^-$ from charge$) = 32$ |
| (c) | $SiO_3^{?-}$ | $4 + 3(6) + (2\ e^-$ from charge$) = 24$ |
| (d) | $CNS^{?-}$ | $4 + 5 + 6 + (2\ e^-$ from charge$) = 17$ |

The polyatomic anions with an even number of valence electrons are (b) and (c). Thus, $S_2O_3^{2-}$ and $SiO_3^{2-}$ are the polyatomic anions that have a 2– charge.

57.

| Ions | $Ag^+$ | $Hg_2^{2+}$ | $Al^{3+}$ | $Sn^{4+}$ |
|---|---|---|---|---|
| $Br^-$ | $AgBr$<br>silver bromide | $Hg_2Br_2$<br>mercury(I) bromide | $AlBr_3$<br>aluminum bromide | $SnBr_4$<br>tin(IV) bromide |
| $F^-$ | $AgF$<br>silver fluoride | $Hg_2F_2$<br>mercury(I) fluoride | $AlF_3$<br>aluminum fluoride | $SnF_4$<br>tin(IV) fluoride |
| $O^{2-}$ | $Ag_2O$<br>silver oxide | $Hg_2O$<br>mercury(I) oxide | $Al_2O_3$<br>aluminum oxide | $SnO_2$<br>tin(IV) oxide |
| $N^{3-}$ | $Ag_3N$<br>silver nitride | $(Hg_2)_3N_2$<br>mercury(I) nitride | $AlN$<br>aluminum nitride | $Sn_3N_4$<br>tin(IV) nitride |

59.

| Ions | $K^+$ | $Cd^{2+}$ | $Cr^{3+}$ | $Bi^{3+}$ |
|---|---|---|---|---|
| $CrO_4^{2-}$ | $K_2CrO_4$<br>potassium chromate | $CdCrO_4$<br>cadmium chromate | $Cr_2(CrO_4)_3$<br>chromium(III) chromate | $Bi_2(CrO_4)_3$<br>bismuth chromate |
| $MnO_4^-$ | $KMnO_4$<br>potassium permanganate | $Cd(MnO_4)_2$<br>cadmium permanganate | $Cr(MnO_4)_3$<br>chromium(III) permanganate | $Bi(MnO_4)_3$<br>bismuth permanganate |
| $ClO_4^-$ | $KClO_4$<br>potassium perchlorate | $Cd(ClO_4)_2$<br>cadmium perchlorate | $Cr(ClO_4)_3$<br>chromium(III) perchlorate | $Bi(ClO_4)_3$<br>bismuth perchlorate |
| $SO_3^{2-}$ | $K_2SO_3$<br>potassium sulfite | $CdSO_3$<br>cadmium sulfite | $Cr_2(SO_3)_3$<br>chromium(III) sulfite | $Bi_2(SO_3)_3$<br>bismuth sulfite |

61.

| | Compound | Suffix Ending | | Compound | Suffix Ending |
|---|---|---|---|---|---|
| (a) | $Na_2S$ | -ide | (b) | $H_2S_{(aq)}$ | -ic acid |
| (c) | $Na_2SO_3$ | -ite | (d) | $H_2SO_{3(aq)}$ | -ous acid |
| (e) | $Na_2SO_4$ | -ate | (f) | $H_2SO_{4(aq)}$ | -ic acid |

63.

| | Chemical Name | Common Name | Formula |
|---|---|---|---|
| (a) | hydrogen dioxide | hydrogen peroxide | $H_2O_2$ |
| (b) | sodium hypochlorite | bleach | $NaClO$ |
| (c) | sodium hydroxide | caustic soda | $NaOH$ |
| (d) | sodium bicarbonate | baking soda | $NaHCO_3$ |

65.

| | Binary Compound | Systematic Name |
|---|---|---|
| (a) | $BF_3$ | boron trifluoride |
| (b) | $SiCl_4$ | silicon tetrachloride |
| (c) | $As_2O_5$ | diarsenic pentaoxide |
| (d) | $Sb_2O_3$ | diantimony trioxide |

67.

| Name | Formula |
|---|---|
| calcium acetate | $Ca(C_2H_3O_2)_2$ |

69. Sodium acid salt $NaHC_2O_4$

# CHAPTER 7 Chemical Formula Calculations

## Section 7.1 *Avogadro's Number*

1.

| | Element | Average Mass | | Element | Average Mass |
|---|---|---|---|---|---|
| (a) | H | 1.0 amu | (b) | Li | 6.9 amu |
| (c) | C | 12.0 amu | (d) | P | 31.0 amu |
| (e) | Ca | 40.1 amu | (f) | Zn | 65.4 amu |
| (g) | As | 74.9 amu | (h) | U | 238.0 amu |

## Section 7.2 *The Mole Concept*

3. (a) 1 mole Mn atoms $= 6.02 \times 10^{23}$ atoms Mn
   (b) 1 mole $NO_2$ molecules $= 6.02 \times 10^{23}$ molecules $NO_2$
   (c) 1 mole $Mn(NO_3)_2$ formula units
   $= 6.02 \times 10^{23}$ formula units $Mn(NO_3)_2$
   (d) 1 mole $MnO_4^-$ ions $= 6.02 \times 10^{23}$ ions $MnO_4^-$

5. (a) $0.335 \cancel{\text{mol Ti}} \times \dfrac{6.02 \times 10^{23} \text{ atoms Ti}}{1 \cancel{\text{mol Ti}}} = 2.02 \times 10^{23}$ atoms Ti

   (b) $0.112 \cancel{\text{mol CO}_2} \times \dfrac{6.02 \times 10^{23} \text{ molecules CO}_2}{1 \cancel{\text{mol CO}_2}}$
   $= 6.74 \times 10^{22}$ molecules $CO_2$

   (c) $1.94 \cancel{\text{mol ZnCl}_2} \times \dfrac{6.02 \times 10^{23} \text{ formula units ZnCl}_2}{1 \cancel{\text{mol ZnCl}_2}}$
   $= 1.17 \times 10^{24}$ formula units $ZnCl_2$

   (d) $0.335 \cancel{\text{mol NO}_2^-} \times \dfrac{6.02 \times 10^{23} \text{ ions NO}_2^-}{1 \cancel{\text{mol NO}_2^-}} = 2.02 \times 10^{23}$ ions $NO_2^-$

7. (a) $4.15 \times 10^{22}$ ~~atoms Fe~~ $\times \dfrac{1 \text{ mol Fe}}{6.02 \times 10^{23} \text{ atoms Fe}} = 0.0689$ mol Fe

(b) $3.31 \times 10^{21}$ ~~molecules $Br_2$~~ $\times \dfrac{1 \text{ mol } Br_2}{6.02 \times 10^{23} \text{ molecules } Br_2}$

$= 5.50 \times 10^{-3}$ mol $Br_2$

(c) $4.19 \times 10^{20}$ ~~formula units $Cd(NO_3)_2$~~ $\times$

$\dfrac{1 \text{ mol } Cd(NO_3)_2}{6.02 \times 10^{23} \text{ formula units } Cd(NO_3)_2} = 6.96 \times 10^{-4}$ mol $Cd(NO_3)_2$

(d) $8.12 \times 10^{23}$ ~~ions $C_2H_3O_2^-$~~ $\times \dfrac{1 \text{ mol } C_2H_3O_2^-}{6.02 \times 10^{23} \text{ ions } C_2H_3O_2^-}$

$= 1.35$ mol $C_2H_3O_2^-$

**Section 7.3** *Molar Mass*

9.

| | Element | Molar Mass | | Element | Molar Mass |
|---|---|---|---|---|---|
| (a) | Hg | 200.6 g/mol | (b) | Si | 28.1 g/mol |
| (c) | Cu | 63.5 g/mol | (d) | Se | 79.0 g/mol |

11.

| | Ionic Compound | Molar Mass |
|---|---|---|
| (a) | $BaF_2$ | 137.3 g + 2(19.0) g = 175.3 g/mol |
| (b) | $K_2S$ | 2(39.1) g + 32.1 g = 110.3 g/mol |
| (c) | $Fe(C_2H_3O_2)_3$ | 55.8 g + 6(12.0) g + 9(1.0) g + 6(16.0) g = 232.8 g/mol |
| (d) | $Sr_3(PO_4)_2$ | 3(87.6) g + 2(31.0) g + 8(16.0) g = 452.8 g/mol |

**Section 7.4** *Mole Calculations*

13. (a) MM of Hg = 200.6 g/mol

$2.95 \times 10^{23}$ ~~atoms Hg~~ $\times \dfrac{1 \text{ mol Hg}}{6.02 \times 10^{23} \text{ atoms Hg}} \times \dfrac{200.6 \text{ g Hg}}{1 \text{ mol Hg}}$

$= 98.3$ g Hg

(b) MM of $N_2$ = 2(14.0) g = 28.0 g/mol

$1.16 \times 10^{22}$ ~~molecules $N_2$~~ $\times \dfrac{1 \text{ mol } N_2}{6.02 \times 10^{23} \text{ molecules } N_2} \times \dfrac{28.0 \text{ g } N_2}{1 \text{ mol } N_2}$

$= 0.540$ g $N_2$

(c) MM of $BaCl_2$ = 137.3 g + 2(35.5) g = 208.3 g/mol

$5.05 \times 10^{21}$ ~~formula units $BaCl_2$~~ $\times$

$\dfrac{1 \text{ mol } BaCl_2}{6.02 \times 10^{23} \text{ formula units } BaCl_2} \times \dfrac{208.3 \text{ g } BaCl_2}{1 \text{ mol } BaCl_2} = 1.75$ g $BaCl_2$

15. (a) MM of K = 39.1 g/mol

$$1.50\ \cancel{\text{g K}} \times \frac{1\ \cancel{\text{mol K}}}{39.1\ \cancel{\text{g K}}} \times \frac{6.02 \times 10^{23}\ \text{atoms K}}{1\ \cancel{\text{mol K}}} = 2.31 \times 10^{22}\ \text{atoms K}$$

(b) MM of $O_2$ = 2(16.0) g = 32.0 g/mol

$$0.470\ \cancel{\text{g O}_2} \times \frac{1\ \cancel{\text{mol O}_2}}{32.0\ \cancel{\text{g O}_2}} \times \frac{6.02 \times 10^{23}\ \text{molecules O}_2}{1\ \cancel{\text{mol O}_2}}$$

$$= 8.84 \times 10^{21}\ \text{molecules O}_2$$

(c) MM of $CsClO_3$ = 132.9 g + 35.5 g + 3(16.0) g = 216.4 g/mol

$$20.36\ \cancel{\text{g CsClO}_3} \times \frac{1\ \cancel{\text{mol CsClO}_3}}{216.4\ \cancel{\text{g CsClO}_3}} \times$$

$$\frac{6.02 \times 10^{23}\ \text{formula units CsClO}_3}{1\ \cancel{\text{mol CsClO}_3}} = 5.66 \times 10^{22}\ \text{formula units CsClO}_3$$

17. (a) $$\frac{9.0\ \text{g Be}}{1\ \cancel{\text{mol Be}}} \times \frac{1\ \cancel{\text{mol Be}}}{6.02 \times 10^{23}\ \text{atoms Be}} = 1.50 \times 10^{-23}\ \text{g/atom}$$

(b) $$\frac{23.0\ \text{g Na}}{1\ \cancel{\text{mol Na}}} \times \frac{1\ \cancel{\text{mol Na}}}{6.02 \times 10^{23}\ \text{atoms Na}} = 3.82 \times 10^{-23}\ \text{g/atom}$$

(c) $$\frac{58.9\ \text{g Co}}{1\ \cancel{\text{mol Co}}} \times \frac{1\ \cancel{\text{mol Co}}}{6.02 \times 10^{23}\ \text{atoms Co}} = 9.79 \times 10^{-23}\ \text{g/atom}$$

(d) $$\frac{74.9\ \text{g As}}{1\ \cancel{\text{mol As}}} \times \frac{1\ \cancel{\text{mol As}}}{6.02 \times 10^{23}\ \text{atoms As}} = 1.24 \times 10^{-22}\ \text{g/atom}$$

**Section 7.5** *Percentage Composition*

19. MM of $C_7H_6O_3$ = 7(12.0) g C + 6(1.0) g H + 3(16.0) g O
= 84.0 g C + 6.0 g H + 48.0 g O = 138.0 g $C_7H_6O_3$

$$\frac{84.0\ \text{g C}}{138.0\ \text{g C}_7\text{H}_6\text{O}_3} \times 100 = 60.9\%\ \text{C}$$

$$\frac{6.0\ \text{g H}}{138.0\ \text{g C}_7\text{H}_6\text{O}_3} \times 100 = 4.4\%\ \text{H}$$

$$\frac{48.0\ \text{g O}}{138.0\ \text{g C}_7\text{H}_6\text{O}_3} \times 100 = 34.7\%\ \text{O}$$

21. MM of $C_{17}H_{21}NO_4$ = 17(12.0) g C + 21(1.0) g H + 14.0 g N + 4(16.0) g O
= 204.0 g C + 21.0 g H + 14.0 g N + 64.0 g O
= 303.0 g $C_{17}H_{21}NO_4$

$$\frac{204.0 \text{ g C}}{303.0 \text{ g } C_{17}H_{21}NO_4} \times 100 = 67.3\% \text{ C}$$
$$\frac{21.0 \text{ g H}}{303.0 \text{ g } C_{17}H_{21}NO_4} \times 100 = 6.9\% \text{ H}$$
$$\frac{14.0 \text{ g N}}{303.0 \text{ g } C_{17}H_{21}NO_4} \times 100 = 4.6\% \text{ N}$$
$$\frac{64.0 \text{ g O}}{303.0 \text{ g } C_{17}H_{21}NO_4} \times 100 = 21.1\% \text{ O}$$

23. MM of $C_4H_8SCl_2$ = 4(12.0) g C + 8(1.0) g H + 32.1 g S + 2(35.5) g Cl
= 48.0 g C + 8.0 g H + 32.1 g S + 71.0 g Cl
= 159.1 g $C_4H_8SCl_2$

$$\frac{48.0 \text{ g C}}{159.1 \text{ g } C_4H_8SCl_2} \times 100 = 30.2\% \text{ C}$$
$$\frac{8.0 \text{ g H}}{159.1 \text{ g } C_4H_8SCl_2} \times 100 = 5.0\% \text{ H}$$
$$\frac{32.1 \text{ g S}}{159.1 \text{ g } C_4H_8SCl_2} \times 100 = 20.2\% \text{ S}$$
$$\frac{71.0 \text{ g Cl}}{159.1 \text{ g } C_4H_8SCl_2} \times 100 = 44.6\% \text{ Cl}$$

25. MM of $NaC_5H_8NO_4$ = 23.0 g Na + 5(12.0) g C + 8(1.0) g H + 14.0 g N + 4(16.0) g O
= 23.0 g Na + 60.0 g C + 8.0 g H + 14.0 g N + 64.0 g O
= 169.0 g $NaC_5H_8NO_4$

$$\frac{23.0 \text{ g Na}}{169.0 \text{ g } NaC_5H_8NO_4} \times 100 = 13.6\% \text{ Na}$$
$$\frac{60.0 \text{ g C}}{169.0 \text{ g } NaC_5H_8NO_4} \times 100 = 35.5\% \text{ C}$$
$$\frac{8.0 \text{ g H}}{169.0 \text{ g } NaC_5H_8NO_4} \times 100 = 4.7\% \text{ H}$$
$$\frac{14.0 \text{ g N}}{169.0 \text{ g } NaC_5H_8NO_4} \times 100 = 8.3\% \text{ N}$$
$$\frac{64.0 \text{ g O}}{169.0 \text{ g } NaC_5H_8NO_4} \times 100 = 37.9\% \text{ O}$$

27. $0.500\ \cancel{g\ Sn} \times \frac{1\ mol\ Sn}{118.7\ \cancel{g\ Sn}} = 0.00421\ mol\ Sn$

$0.635\ g\ Sn_xO_y - 0.500\ g\ Sn = 0.135\ g\ O$

$0.135\ \cancel{g\ O} \times \frac{1\ mol\ O}{16.0\ \cancel{g\ O}} = 0.00844\ mol\ O$

$Sn_{\frac{0.00421}{0.00421}}O_{\frac{0.00844}{0.00421}} = Sn_{1.00}O_{2.00}$

The empirical formula is $SnO_2$.

29. $1.435\ \cancel{g\ Hg} \times \frac{1\ mol\ Hg}{200.6\ \cancel{g\ Hg}} = 0.007154\ mol\ Hg$

$1.550\ g\ Hg_xO_y - 1.435\ g\ Hg = 0.115\ g\ O$

$0.115\ \cancel{g\ O} \times \frac{1\ mol\ O}{16.0\ \cancel{g\ O}} = 0.00719\ mol\ O$

$Hg_{\frac{0.007154}{0.007154}}O_{\frac{0.00719}{0.007154}} = Hg_{1.000}O_{1.01}$

The empirical formula is HgO.

31. $1.115\ \cancel{g\ Co} \times \frac{1\ mol\ Co}{58.9\ \cancel{g\ Co}} = 0.0189\ mol\ Co$

$2.025\ g\ Co_xS_y - 1.115\ g\ Co = 0.910\ g\ S$

$0.910\ \cancel{g\ S} \times \frac{1\ mol\ S}{32.1\ \cancel{g\ S}} = 0.0283\ mol\ S$

$Co_{\frac{0.0189}{0.0189}}S_{\frac{0.0283}{0.0189}} = Co_{1.00}S_{1.50}$

The empirical formula is $Co_2S_3$.

33.* (a) $59.1\ \cancel{g\ Mn} \times \frac{1\ mol\ Mn}{54.9\ \cancel{g\ Mn}} = 1.08\ mol\ Mn$

$40.9\ \cancel{g\ F} \times \frac{1\ mol\ F}{19.0\ \cancel{g\ F}} = 2.15\ mol\ F$

$Mn_{\frac{1.08}{1.08}}F_{\frac{2.15}{1.08}} = Mn_{1.00}F_{1.99}$

The empirical formula is $MnF_2$.

* continued on next page

33. (b) $64.1\ \cancel{g\ Cu} \times \frac{1\ mol\ Cu}{63.5\ \cancel{g\ Cu}} = 1.01\ mol\ Cu$

$35.9\ \cancel{g\ Cl} \times \frac{1\ mol\ Cl}{35.5\ \cancel{g\ Cl}} = 1.01\ mol\ Cl$

$\frac{Cu_{1.01}}{1.01}\frac{Cl_{1.01}}{1.01} = Cu_{1.00}Cl_{1.00}$

The empirical formula is CuCl.

(c) $42.6\ \cancel{g\ Sn} \times \frac{1\ mol\ Sn}{118.7\ \cancel{g\ Sn}} = 0.359\ mol\ Sn$

$57.4\ \cancel{g\ Br} \times \frac{1\ mol\ Br}{79.9\ \cancel{g\ Br}} = 0.718\ mol\ Br$

$\frac{Sn_{0.359}}{0.359}\frac{Br_{0.718}}{0.359} = Sn_{1.00}Br_{2.00}$

The empirical formula is $SnBr_2$.

(d) $61.7\ \cancel{g\ Tl} \times \frac{1\ mol\ Tl}{204.4\ \cancel{g\ Tl}} = 0.302\ mol\ Tl$

$38.3\ \cancel{g\ I} \times \frac{1\ mol\ I}{126.9\ \cancel{g\ I}} = 0.302\ mol\ I$

$\frac{Tl_{0.302}}{0.302}\frac{I_{0.302}}{0.302} = Tl_{1.00}I_{1.00}$

The empirical formula is TlI.

35. $18.25\ \cancel{g\ C} \times \frac{1\ mol\ C}{12.0\ \cancel{g\ C}} = 1.52\ mol\ C$

$0.77\ \cancel{g\ H} \times \frac{1\ mol\ H}{1.0\ \cancel{g\ H}} = 0.77\ mol\ H$

$80.99\ \cancel{g\ Cl} \times \frac{1\ mol\ Cl}{35.5\ \cancel{g\ Cl}} = 2.28\ mol\ Cl$

$\frac{C_{1.52}}{0.77}\frac{H_{0.77}}{0.77}\frac{Cl_{2.28}}{0.77} = C_{2.0}H_{1.0}Cl_{3.0}$

The empirical formula is $C_2HCl_3$.

**Section 7.7** *Molecular Formula*

37. MM of $C_9H_8O_4$ $= 9(12.0)\ g\ C + 8(1.0)\ g\ H + 4(16.0)\ g\ O$
$= 180.0\ g/mol\ C_9H_8O_4$

aspirin: $\frac{(C_9H_8O_4)_n}{C_9H_8O_4} = \frac{180\ g/mol}{180.0\ g/mol}$ $n \approx 1$

The molecular formula of aspirin is $(C_9H_8O_4)_1$ or $C_9H_8O_4$.

39. MM of $C_3H_5O_2$ = 3(12.0) g C + 5(1.0) g H + 2(16.0) g O
= 73.0 g/mol $C_3H_5O_2$

adipic acid: $\frac{(C_3H_5O_2)_n}{C_3H_5O_2} = \frac{147\text{ g/mol}}{73.0\text{ g/mol}}$ $n \approx 2$

The molecular formula of adipic acid is $(C_3H_5O_2)_2$ or $C_6H_{10}O_4$.

41. Empirical Formula

$38.7\ \cancel{\text{g C}} \times \frac{1\text{ mol C}}{12.0\ \cancel{\text{g C}}} = 3.23\text{ mol C}$

$9.74\ \cancel{\text{g H}} \times \frac{1\text{ mol H}}{1.0\ \cancel{\text{g H}}} = 9.74\text{ mol H}$

$51.6\ \cancel{\text{g O}} \times \frac{1\text{ mol O}}{16.0\ \cancel{\text{g O}}} = 3.23\text{ mol O}$

$\frac{C_{3.23}}{3.23}\frac{H_{9.74}}{3.23}\frac{O_{3.23}}{3.23} = C_{1.00}H_{3.02}O_{1.00}$

The empirical formula is $CH_3O$.

Molecular Formula

MM of $CH_3O$ = 12.0 g C + 3(1.0) g H + 16.0 g O = 31.0 g/mol $CH_3O$

ethylene glycol: $\frac{(CH_3O)_n}{CH_3O} = \frac{62\text{ g/mol}}{31.0\text{ g/mol}}$ $n \approx 2$

The molecular formula of ethylene glycol is $(CH_3O)_2$ or $C_2H_6O_2$.

43.* Empirical Formula

$24.8\ \cancel{\text{g C}} \times \frac{1\text{ mol C}}{12.0\ \cancel{\text{g C}}} = 2.07\text{ mol C}$

$2.08\ \cancel{\text{g H}} \times \frac{1\text{ mol H}}{1.0\ \cancel{\text{g H}}} = 2.08\text{ mol H}$

$73.1\ \cancel{\text{g Cl}} \times \frac{1\text{ mol Cl}}{35.5\ \cancel{\text{g Cl}}} = 2.06\text{ mol Cl}$

$\frac{C_{2.07}}{2.06}\frac{H_{2.08}}{2.06}\frac{Cl_{2.06}}{2.06} = C_{1.00}H_{1.01}Cl_{1.00}$

The empirical formula is CHCl.

---

* continued on next page

43. Molecular Formula

MM of CHCl = 12.0 g C + 1.0 g H + 35.5 g Cl = 48.5 g/mol CHCl

lindane: $\frac{(CHCl)_n}{CHCl} = \frac{290 \text{ g/mol}}{48.5 \text{ g/mol}}$ $\quad n \approx 6$

The molecular formula of lindane is $(CHCl)_6$ or $C_6H_6Cl_6$.

45. Empirical Formula

$74.0 \cancel{\text{g C}} \times \frac{1 \text{ mol C}}{12.0 \cancel{\text{g C}}} = 6.17 \text{ mol C}$

$8.70 \cancel{\text{g H}} \times \frac{1 \text{ mol H}}{1.0 \cancel{\text{g H}}} = 8.70 \text{ mol H}$

$17.3 \cancel{\text{g N}} \times \frac{1 \text{ mol N}}{14.0 \cancel{\text{g N}}} = 1.24 \text{ mol N}$

$\frac{C_{6.17}}{1.24}\frac{H_{8.70}}{1.24}\frac{N_{1.24}}{1.24} = C_{4.98}H_{7.02}N_{1.00}$

The empirical formula is $C_5H_7N$.

Molecular Formula

MM of $C_5H_7N$ = 5(12.0) g C + 7(1.0) g H + 14.0 g N = 81.0 g/mol CHCl

nicotine: $\frac{(C_5H_7N)_n}{C_5H_7N} = \frac{160 \text{ g/mol}}{81.0 \text{ g/mol}}$ $\quad n \approx 2$

The molecular formula of nicotine is $(C_5H_7N)_2$ or $C_{10}H_{14}N_2$.

**Section 7.8** *Molar Volume*

47. Standard conditions: 0°C and 1 atmosphere (1 atm)

49.

| Gas | Molecules | Mass | Vol @ STP |
|---|---|---|---|
| fluorine, $F_2$ | $6.02 \times 10^{23}$ | 38.0 g | 22.4 L |
| hydrogen fluoride, HF | $6.02 \times 10^{23}$ | 20.0 g | 22.4 L |
| silicon tetrafluoride, $SiF_4$ | $6.02 \times 10^{23}$ | 104 g | 22.4 L |
| oxygen difluoride, $OF_2$ | $6.02 \times 10^{23}$ | 54.0 g | 22.4 L |

51. (a) MM of Ne = 20.2 g/mol Ne

Density of Ne (at STP): $\frac{20.2 \text{ g}}{\cancel{\text{mol}}} \times \frac{1 \cancel{\text{mol}}}{22.4 \text{ L}} = 0.902 \text{ g/L}$

(b) MM of $Cl_2$ = 2(35.5) g Cl = 71.0 g/mol $Cl_2$

Density of $Cl_2$ (at STP): $\frac{71.0 \text{ g}}{\cancel{\text{mol}}} \times \frac{1 \cancel{\text{mol}}}{22.4 \text{ L}} = 3.17 \text{ g/L}$

(c) MM of $NO_2$ = 14.0 g N + 2(16.0) g O = 46.0 g/mol $NO_2$

Density of $NO_2$ (at STP): $\frac{46.0 \text{ g}}{\cancel{\text{mol}}} \times \frac{1 \cancel{\text{mol}}}{22.4 \text{ L}} = 2.05 \text{ g/L}$

(d) MM of HI = 1.0 g H + 126.9 g I = 127.9 g/mol HI

Density of HI (at STP): $\frac{127.9 \text{ g}}{\cancel{\text{mol}}} \times \frac{1 \cancel{\text{mol}}}{22.4 \text{ L}} = 5.71 \text{ g/L}$

53. (a) MM of ethane: $\frac{22.4 \cancel{\text{L}}}{\text{mol}} \times \frac{1.34 \text{ g}}{\cancel{\text{L}}} = 30.0 \text{ g/mol}$

(b) MM of diborane: $\frac{22.4 \cancel{\text{L}}}{\text{mol}} \times \frac{1.23 \text{ g}}{\cancel{\text{L}}} = 27.6 \text{ g/mol}$

(c) MM of Freon-12: $\frac{22.4 \cancel{\text{L}}}{\text{mol}} \times \frac{5.40 \text{ g}}{\cancel{\text{L}}} = 121 \text{ g/mol}$

(d) MM of nitrous oxide: $\frac{22.4 \cancel{\text{L}}}{\text{mol}} \times \frac{2.05 \text{ g}}{\cancel{\text{L}}} = 45.9 \text{ g/mol}$

**Section 7.9** *Mole Calculations II*

55. (a) MM of $H_2S$ = 2(1.0) g H + 32.1 g S = 34.1 g/mol $H_2S$

$1.05 \cancel{\text{L H}_2\text{S}} \times \frac{1 \cancel{\text{mol H}_2\text{S}}}{22.4 \cancel{\text{L H}_2\text{S}}} \times \frac{34.1 \text{ g H}_2\text{S}}{1 \cancel{\text{mol H}_2\text{S}}} = 1.60 \text{ g H}_2\text{S}$

(b) MM of $N_2O_3$ = 2(14.0) g N + 3(16.0) g O = 76.0 g/mol $N_2O_3$

$5.33 \cancel{\text{L N}_2\text{O}_3} \times \frac{1 \cancel{\text{mol N}_2\text{O}_3}}{22.4 \cancel{\text{L N}_2\text{O}_3}} \times \frac{76.0 \text{ g N}_2\text{O}_3}{1 \cancel{\text{mol N}_2\text{O}_3}} = 18.1 \text{ g N}_2\text{O}_3$

(c) MM of $Cl_2O$ = 2(35.5) g Cl + 16.0 g O = 87.0 g/mol $Cl_2O$

$75.0 \cancel{\text{mL Cl}_2\text{O}} \times \frac{1 \cancel{\text{L Cl}_2\text{O}}}{1000 \cancel{\text{mL Cl}_2\text{O}}} \times \frac{1 \cancel{\text{mol Cl}_2\text{O}}}{22.4 \cancel{\text{L Cl}_2\text{O}}} \times \frac{87.0 \text{ g Cl}_2\text{O}}{1 \cancel{\text{mol Cl}_2\text{O}}}$

$= 0.291 \text{ g Cl}_2\text{O}$

57. (a) $10.0\ \cancel{\text{mL H}_2} \times \dfrac{1\ \cancel{\text{L H}_2}}{1000\ \cancel{\text{mL H}_2}} \times \dfrac{1\ \cancel{\text{mol H}_2}}{22.4\ \cancel{\text{L H}_2}} \times \dfrac{6.02 \times 10^{23}\ \text{molecules H}_2}{1\ \cancel{\text{mol H}_2}}$

$= 2.69 \times 10^{20}\ \text{molecules H}_2$

(b) $70.5\ \cancel{\text{mL NH}_3} \times \dfrac{1\ \cancel{\text{L NH}_3}}{1000\ \cancel{\text{mL NH}_3}} \times \dfrac{1\ \cancel{\text{mol NH}_3}}{22.4\ \cancel{\text{L NH}_3}} \times$

$\dfrac{6.02 \times 10^{23}\ \text{molecules NH}_3}{1\ \cancel{\text{mol NH}_3}} = 1.89 \times 10^{21}\ \text{molecules NH}_3$

(c) $1.00\ \cancel{\text{L CO}_2} \times \dfrac{1\ \cancel{\text{mol CO}_2}}{22.4\ \cancel{\text{L CO}_2}} \times \dfrac{6.02 \times 10^{23}\ \text{molecules CO}_2}{1\ \cancel{\text{mol CO}_2}}$

$= 2.69 \times 10^{22}\ \text{molecules CO}_2$

59.

| Gas | Molecules | Atoms | Mass | Vol @ STP |
|---|---|---|---|---|
| $N_2$ | $1.35 \times 10^{23}$ | $2.71 \times 10^{23}$ | 6.31 g | 5.04 L |
| $NO_2$ | $1.35 \times 10^{23}$ | $4.06 \times 10^{23}$ | 10.4 g | 5.04 L |
| NO | $1.35 \times 10^{23}$ | $2.71 \times 10^{23}$ | 6.75 g | 5.04 L |
| $N_2O_4$ | $1.35 \times 10^{23}$ | $8.13 \times 10^{23}$ | 20.7 g | 5.04 L |

**General Exercises**

61. 1 faraday = 1 mole of electrons = $6.02 \times 10^{23}$ electrons
1 einstein = 1 mole of photons = $6.02 \times 10^{23}$ photons

63. $5.0\ \cancel{\text{g Ni}} \times \dfrac{1\ \cancel{\text{mol Ni}}}{58.7\ \cancel{\text{g Ni}}} \times \dfrac{6.02 \times 10^{23}\ \text{atoms Ni}}{1\ \cancel{\text{mol Ni}}} = 5.1 \times 10^{22}\ \text{atoms Ni}$

$5.1 \times 10^{22}$ atoms Ni > $1 \times 10^{15}$ red blood cells

Thus, the number of Ni atoms in a 5.0 g Ni coin is 51,000,000 times greater than the number of red blood cells in 50,000 people.

65. $1\ \cancel{\text{mol furry moles}} \times \dfrac{6.02 \times 10^{23}\ \cancel{\text{furry moles}}}{1\ \cancel{\text{mol furry moles}}} \times \dfrac{100\ \cancel{\text{g}}}{1\ \cancel{\text{furry moles}}} \times \dfrac{1\ \text{kg}}{1000\ \cancel{\text{g}}}$

$= 6 \times 10^{22}\ \text{kg}$

$6 \times 10^{24}$ kg (earth) > $6 \times 10^{22}$ kg (1 mol furry moles)

Therefore, the earth weighs about 100 times more than the mole of moles!

67. $0.500\ \cancel{\text{g Ga}} \times \dfrac{1\ \text{mol Ga}}{69.7\ \cancel{\text{g Ga}}} = 0.00717\ \text{mol Ga}$

$0.672\ \text{g}\ Ga_xO_y - 0.500\ \text{g Ga} = 0.172\ \text{g O}$

$0.172\ \cancel{\text{g O}} \times \dfrac{1\ \text{mol O}}{16.0\ \cancel{\text{g O}}} = 0.0108\ \text{mol O}$

$\dfrac{Ga_{0.00717}}{0.00717}\dfrac{O_{0.0108}}{0.00717} = Ga_{1.00}O_{1.51}$

The empirical formula is $Ga_2O_3$.

69. $1\ \cancel{\text{molecule } H_2O} \times \dfrac{1\ \cancel{\text{mol } H_2O}}{6.02 \times 10^{23}\ \cancel{\text{molecules } H_2O}} \times \dfrac{18.0\ \cancel{\text{g } H_2O}}{1\ \cancel{\text{mol } H_2O}} \times \dfrac{1\ \text{cm}^3\ H_2O}{1\ \cancel{\text{g } H_2O}} = 2.99 \times 10^{-23}\ \text{cm}^3\ H_2O$

71. MM of $C_{12}H_{22}O_{11}$ $= 12(12.0)\ \text{g C} + 22(1.0)\ \text{g H} + 11(16.0)\ \text{g O}$
$= 342.0\ \text{g/mol}\ C_{12}H_{22}O_{11}$

$1.00\ \cancel{\text{g } C_{12}H_{22}O_{11}} \times \dfrac{1\ \cancel{\text{mol } C_{12}H_{22}O_{11}}}{342.0\ \cancel{\text{g } C_{12}H_{22}O_{11}}} \times \dfrac{12\ \cancel{\text{mol C}}}{1\ \cancel{\text{mol } C_{12}H_{22}O_{11}}} \times \dfrac{6.02 \times 10^{23}\ \text{atoms C}}{1\ \cancel{\text{mol C}}} = 2.11 \times 10^{22}\ \text{atoms C}$

73. $1\ \cancel{\text{molecule vitamin K}} \times \dfrac{1\ \cancel{\text{mol vitamin K}}}{6.02 \times 10^{23}\ \cancel{\text{molecules vitamin K}}} \times \dfrac{173\ \cancel{\text{g vitamin K}}}{1\ \cancel{\text{mol vitamin K}}} \times \dfrac{76.3\ \cancel{\text{g C}}}{100\ \cancel{\text{g vitamin K}}} \times \dfrac{1\ \cancel{\text{mol C}}}{12.0\ \cancel{\text{g C}}} \times \dfrac{6.02 \times 10^{23}\ \text{atoms C}}{1\ \cancel{\text{mol C}}} = 11\ \text{atoms C}$

(Note: Since the number of atoms must be a whole number, the answer is 11 atoms not 11.0 atoms.)

75. Volume of Cu atom in cm$^3$: $\dfrac{1\ \text{atom Cu}}{0.0118\ \cancel{\text{nm}^3}} \times \left(\dfrac{10^9\ \cancel{\text{nm}}}{1\ \cancel{\text{m}}}\right)^3 \times \left(\dfrac{1\ \cancel{\text{m}}}{100\ \text{cm}}\right)^3 = \dfrac{8.47 \times 10^{22}\ \text{atoms Cu}}{\cancel{\text{cm}^3}}$

$\dfrac{8.47 \times 10^{22}\ \text{atoms Cu}}{\cancel{\text{cm}^3}} \times \dfrac{\cancel{\text{cm}^3}}{8.92\ \cancel{\text{g Cu}}} \times \dfrac{63.5\ \cancel{\text{g Cu}}}{1\ \text{mol Cu}} = \dfrac{6.03 \times 10^{23}\ \text{atoms Cu}}{\text{mol Cu}}$

# Writing Chemical Equations

CHAPTER

**8**

## Section 8.1 *Evidence for Chemical Reactions*

1. (a) The production of a gas is evidence of a chemical reaction.
   (b) The extinguishing of a flaming splint is evidence of a chemical reaction.
   (c) The formation of a precipitate is evidence of a chemical reaction.
   (d) A permanent color change is evidence of a chemical reaction.
   (e) A heat energy change is evidence of a chemical reaction.

## Section 8.2 *Writing Chemical Equations*

3. (a) $Sn_{(s)} + O_{2(g)} \rightarrow SnO_{2(s)}$
   (b) $ZnCO_{3(s)} \rightarrow ZnO_{(s)} + CO_{2(g)}$
   (c) $Mg_{(s)} + Cd(NO_3)_{2(aq)} \rightarrow Mg(NO_3)_{2(aq)} + Cd_{(s)}$
   (d) $LiBr_{(aq)} + AgNO_{3(aq)} \rightarrow AgBr_{(s)} + LiNO_{3(aq)}$
   (e) $HC_2H_3O_{2(aq)} + KOH_{(aq)} \rightarrow KC_2H_3O_{2(aq)} + H_2O_{(l)}$

## Section 8.3 *Balancing Chemical Equations*

5. The statements (b), (c), and (d) are true; (a) and (e) are false.
   (a) Subscripts, not coefficients, are used to balance the elements in each substance.
   (e) If an equation coefficient is a fraction, multiply coefficients to clear the fraction and give whole numbers.

7. (a) $4\,Co_{(s)} + 3\,O_{2(g)} \rightarrow 2\,Co_2O_{3(s)}$
   (b) $2\,CsClO_{3(s)} \rightarrow 2\,CsCl_{(s)} + 3\,O_{2(g)}$
   (c) $Cu_{(s)} + 2\,AgC_2H_3O_{2(aq)} \rightarrow Cu(C_2H_3O_2)_{2(aq)} + 2\,Ag_{(s)}$
   (d) $Pb(NO_3)_{2(aq)} + LiCl_{(aq)} \rightarrow PbCl_{2(s)} + 2\,LiNO_{3(aq)}$
   (e) $3\,H_2SO_{4(aq)} + 2\,Al(OH)_{3(aq)} \rightarrow Al_2(SO_4)_{3(aq)} + 6\,H_2O_{(l)}$

9. (a) $2\ Pb_{(s)} + O_{2(g)} \rightarrow 2\ PbO_{(s)}$
(b) $2\ LiNO_{3(s)} \rightarrow 2\ LiNO_{2(s)} + O_{2(g)}$
(c) $Mg_{(s)} + 2\ HC_2H_3O_{2(aq)} \rightarrow Mg(C_2H_3O_2)_{2(aq)} + H_{2(g)}$
(d) $Hg_2(NO_3)_{2(aq)} + 2\ NaBr_{(aq)} \rightarrow Hg_2Br_{2(s)} + 2\ NaNO_{3(aq)}$
(e) $H_2CO_{3(aq)} + 2\ NH_4OH_{(aq)} \rightarrow (NH_4)_2CO_{3(aq)} + 2\ HOH_{(l)}$

## Section 8.4 *Classifying Chemical Reactions*

11. Refer to reactions in exercise 7.

Classification

(a) combination reaction
(b) decomposition reaction
(c) single-replacement reaction
(d) double-replacement reaction
(e) neutralization reaction

13. Refer to the reactions in exercise 9.

Classification

(a) combination reaction
(b) decomposition reaction
(c) single-replacement reaction
(d) double-replacement reaction
(e) neutralization reaction

## Section 8.5 *Combination Reactions*

General Form: $A + Z \rightarrow AZ$

15. metal + oxygen gas → metal oxide
(a) $4\ Cu_{(s)} + O_{2(g)} \rightarrow 2\ Cu_2O_{(s)}$
(b) $2\ Sn_{(s)} + O_{2(g)} \rightarrow 2\ SnO_{(s)}$
(c) $4\ Fe_{(s)} + 3\ O_{2(g)} \rightarrow 2\ Fe_2O_{3(s)}$
(d) $Ti_{(s)} + O_{2(g)} \rightarrow TiO_{2(s)}$

17. metal + nonmetal → ionic compound
(a) $Hg_{(l)} + Cl_{2(g)} \rightarrow HgCl_{2(s)}$
(b) $3\ Pb_{(s)} + 2\ P_{(s)} \rightarrow Pb_3P_{2(s)}$
(c) $2\ Fe_{(s)} + 3\ F_{2(g)} \rightarrow 2\ FeF_{3(s)}$
(d) $Co_{(s)} + S_{(s)} \rightarrow CoS_{(s)}$

19. metal + oxygen gas → metal oxide

(a) $2\,Cd + O_2 \rightarrow 2\,CdO$

metal + nonmetal → ionic compound

(b) $2\,Al + 3\,Cl_2 \rightarrow 2\,AlCl_3$

(c) $Sr + S \rightarrow SrS$

(d) $3\,Ba + N_2 \rightarrow Ba_3N_2$

21. metal + oxygen gas → metal oxide

(a) $2\,Sr + O_2 \rightarrow 2\,SrO$

metal + nonmetal → ionic compound

(b) $2\,Bi + 3\,S \rightarrow Bi_2S_3$

(c) $2\,Li + Br_2 \rightarrow 2\,LiBr$

(d) $Mg + Cl_2 \rightarrow MgCl_2$

**Section 8.6** *Decomposition Reactions*

General Form: AZ → A + Z

23. metal hydrogen carbonate → metal carbonate + water + carbon dioxide

(a) $2\,AgHCO_{3(s)} \rightarrow Ag_2CO_{3(s)} + H_2O_{(g)} + CO_{2(g)}$

(b) $Cu(HCO_3)_{2(s)} \rightarrow CuCO_{3(s)} + H_2O_{(g)} + CO_{2(g)}$

25. metal carbonate → metal oxide + carbon dioxide

(a) $CuCO_{3(s)} \rightarrow CuO_{(s)} + CO_{2(g)}$

(b) $MnCO_{3(s)} \rightarrow MnO_{(s)} + CO_{2(g)}$

27. oxygen-containing compounds → oxygen gas

(a) $2\,PbO_{2(s)} \rightarrow 2\,PbO_{(s)} + O_{2(g)}$

(b) $2\,AgNO_{3(s)} \rightarrow 2\,AgNO_{2(s)} + O_{2(g)}$

29. metal hydrogen carbonate → metal carbonate + water + carbon dioxide

(a) $2\,KHCO_{3(s)} \rightarrow K_2CO_{3(s)} + H_2O_{(g)} + CO_{2(g)}$

(b) $Zn(HCO_3)_{2(s)} \rightarrow ZnCO_{3(s)} + H_2O_{(g)} + CO_{2(g)}$

metal carbonate → metal oxide + carbon dioxide

(c) $Li_2CO_{3(s)} \rightarrow Li_2O_{(s)} + CO_{2(g)}$

(d) $CdCO_{3(s)} \rightarrow CdO_{(s)} + CO_{2(g)}$

## Section 8.7 *Activity Series*

31.

| | Element | Placed In: | Observation |
|---|---|---|---|
| (a) | Hg | $Fe(NO_3)_{2(aq)}$ | no reaction; Fe > Hg |
| (b) | Zn | $Fe(NO_3)_{2(aq)}$ | reaction; Zn > Fe |
| (c) | Cd | $Fe(NO_3)_{2(aq)}$ | no reaction; Fe > Cd |
| (d) | Mg | $Fe(NO_3)_{2(aq)}$ | reaction; Mg > Fe |

33.

| | Element | Placed In: | Observation |
|---|---|---|---|
| (a) | Ni | $HCl_{(aq)}$ | reaction; Ni > (H) |
| (b) | Zn | $HCl_{(aq)}$ | reaction; Zn > (H) |
| (c) | Ca | $HCl_{(aq)}$ | reaction; Ca > (H) |
| (d) | Al | $HCl_{(aq)}$ | reaction; Al > (H) |

35.

| | Element | Placed In: | Observation |
|---|---|---|---|
| (a) | Li | $H_2O_{(l)}$ | reaction; Li is an active metal |
| (b) | Mg | $H_2O_{(l)}$ | no reaction; Mg is not an active metal |
| (c) | Ca | $H_2O_{(l)}$ | reaction; Ca is an active metal |
| (d) | Al | $H_2O_{(l)}$ | no reaction; Al is not an active metal |

37.

| | Element | Placed In: | Observation |
|---|---|---|---|
| (a) | Mn | $Al(NO_3)_{3(aq)}$ | no reaction; Al > Mn |
| (b) | Fe | $Sn(C_2H_3O_2)_{2(aq)}$ | reaction; Fe > Sn |
| (c) | Co | $HClO_{3(aq)}$ | reaction; Co > (H) |
| (d) | Ag | $H_2SO_{4(aq)}$ | no reaction; (H) > Ag |

## Section 8.8 *Single-Replacement Reactions*

General Form: A + BZ $\rightarrow$ AZ + B

39. metal$_1$ + aqueous solution$_1$ $\rightarrow$ metal$_2$ + aqueous solution$_2$ or *NR*

(a) $Cu_{(s)} + Al(NO_3)_{3(aq)} \rightarrow$ *NR*

(b) $2\,Al_{(s)} + 3\,Cu(NO_3)_{2(aq)} \rightarrow 2\,Al(NO_3)_{3(aq)} + 3\,Cu_{(s)}$

(c) $Cd_{(s)} + FeSO_{4(aq)} \rightarrow$ *NR*

(d) $Fe_{(s)} + CdSO_{4(aq)} \rightarrow FeSO_{4(aq)} + Cd_{(s)}$

41. metal + aqueous acid $\rightarrow$ aqueous solution + hydrogen gas

(a) $Cd_{(s)} + 2\,HCl_{(aq)} \rightarrow CdCl_{2(aq)} + H_{2(g)}$

(b) $Mn_{(s)} + 2\,HC_2H_3O_{2(aq)} \rightarrow Mn(C_2H_3O_2)_{2(aq)} + H_{2(g)}$

(c) $Zn_{(s)} + H_2SO_{4(aq)} \rightarrow ZnSO_{4(aq)} + H_{2(g)}$

(d) $Mg_{(s)} + H_2CO_{3(aq)} \rightarrow MgCO_{3(s)} + H_{2(g)}$

43. metal + water → metal hydroxide + hydrogen gas (if active metal)

metal + water → *NR* (if not an active metal)

(a) $2\ Li_{(s)} + 2\ H_2O_{(l)} \rightarrow 2\ LiOH_{(aq)} + H_{2(g)}$

(b) $Pb_{(s)} + H_2O_{(l)} \rightarrow$ *NR*

(c) $Co_{(s)} + H_2O_{(l)} \rightarrow$ *NR*

(d) $Ca_{(s)} + 2\ H_2O_{(l)} \rightarrow Ca(OH)_{2(aq)} + H_{2(g)}$

45. $metal_1$ + aqueous $solution_1$ → $metal_2$ + aqueous $solution_2$

(a) $Zn_{(s)} + Pb(NO_3)_{2(aq)} \rightarrow Zn(NO_3)_{2(aq)} + Pb_{(s)}$

(b) $Co_{(s)} + NiSO_{4(aq)} \rightarrow CoSO_{4(aq)} + Ni_{(s)}$

(c) $Ni_{(s)} + SnSO_{4(aq)} \rightarrow NiSO_{4(aq)} + Sn_{(s)}$

$metal_1$ + aqueous $solution_1$ → *NR*

(d) $Sn_{(s)} + LiC_2H_3O_{2(aq)} \rightarrow$ *NR*

metal + aqueous acid → aqueous solution + hydrogen gas

(e) $Zn_{(s)} + 2\ HCl_{(aq)} \rightarrow ZnCl_{2(aq)} + H_{2(g)}$

**Section 8.9** *Solubility Rules*

47.

| | Compound | Solubility | | Compound | Solubility |
|---|---|---|---|---|---|
| (a) | $NH_4OH$ | soluble | (b) | $FeSO_4$ | soluble |
| (c) | $K_2CrO_4$ | soluble | (d) | $Pb(C_2H_3O_2)_2$ | soluble |
| (e) | $Al(NO_3)_3$ | soluble | (f) | $Hg_2Cl_2$ | insoluble |
| (g) | $MnI_2$ | soluble | (h) | $HgBr_2$ | soluble |

**Section 8.10** *Double-Replacement Reactions*

General Form: AX + BZ → AZ + BX

49. aqueous $solution_1$ + aqueous $solution_2$ → precipitate + aqueous $solution_3$

(a) $MgSO_{4(aq)} + BaCl_{2(aq)} \rightarrow BaSO_{4(s)} + MgCl_{2(aq)}$

(b) $2\ AlBr_{3(aq)} + 3\ Na_2CO_{3(aq)} \rightarrow Al_2(CO_3)_{3(s)} + 6\ NaBr_{(aq)}$

(c) $3\ NiSO_{4(aq)} + 2\ Li_3PO_{4(aq)} \rightarrow Ni_3(PO_4)_{2(s)} + 3\ Li_2SO_{4(aq)}$

51. aqueous $solution_1$ + aqueous $solution_2$ → precipitate + aqueous $solution_3$

(a) $MnSO_{4(aq)} + 2\ NH_4OH_{(aq)} \rightarrow Mn(OH)_{2(s)} + (NH_4)_2SO_{4(aq)}$

(b) $ZnCl_{2(aq)} + Hg_2(NO_3)_{2(aq)} \rightarrow Hg_2Cl_{2(s)} + Zn(NO_3)_{2(aq)}$

## Section 8.11 *Neutralization Reactions*

General Form: HX + BOH → BX + HOH

53. aqueous acid + aqueous base → aqueous salt + water
    (a) $2\ HCl_{(aq)} + Ca(OH)_{2(aq)} \rightarrow CaCl_{2(aq)} + 2\ HOH_{(l)}$
    (b) $H_2SO_{4(aq)} + 2\ LiOH_{(aq)} \rightarrow Li_2SO_{4(aq)} + 2\ HOH_{(l)}$
    (c) $2\ HNO_{2(aq)} + Sr(OH)_{2(aq)} \rightarrow Sr(NO_2)_{2(aq)} + 2\ HOH_{(l)}$

55. aqueous acid + aqueous base → aqueous salt + water
    (a) $NaOH_{(aq)} + HNO_{3(aq)} \rightarrow NaNO_{3(aq)} + HOH_{(l)}$

    aqueous acid + aqueous base → precipitate + water
    (b) $3\ Ba(OH)_{2(aq)} + 2\ H_3PO_{4(aq)} \rightarrow Ba_3(PO_4)_{2(s)} + 6\ HOH_{(l)}$

## General Exercises

57. (a) $2\ SO_{2(g)} + O_{2(g)} \rightarrow 2\ SO_{3(g)}$
    (b) $F_{2(g)} + 2\ NaBr_{(aq)} \rightarrow Br_{2(l)} + 2\ NaF_{(aq)}$
    (c) $Cl_{2(g)} + H_2O_{(l)} \rightarrow HCl_{(aq)} + HClO_{(aq)}$
    (d) $PCl_{5(s)} + 4\ H_2O_{(l)} \rightarrow H_3PO_{4(aq)} + 5\ HCl_{(aq)}$
    (e) $Sb_2S_{3(s)} + 6\ HCl_{(aq)} \rightarrow 2\ SbCl_{3(aq)} + 3\ H_2S_{(aq)}$

59. (a) $CH_{4(g)} + 2\ O_{2(g)} \rightarrow CO_{2(g)} + 2\ H_2O_{(g)}$
    (b) $2\ CH_4O_{(l)} + 3\ O_{2(g)} \rightarrow 2\ CO_{2(g)} + 4\ H_2O_{(g)}$
    (c) $C_3H_{8(g)} + 5\ O_{2(g)} \rightarrow 3\ CO_{2(g)} + 4\ H_2O_{(g)}$
    (d) $2\ C_3H_8O_{(l)} + 9\ O_{2(g)} \rightarrow 6\ CO_{2(g)} + 8\ H_2O_{(g)}$

61. Deacon method: $4\ HCl_{(g)} + O_{2(g)} \rightarrow 2\ Cl_{2(g)} + 2\ H_2O_{(g)}$

63. Ostwald process:
    (1) $4\ NH_{3(g)} + 5\ O_{2(g)} \rightarrow 4\ NO_{(g)} + 6\ H_2O_{(g)}$
    (2) $2\ NO_{(g)} + O_{2(g)} \rightarrow 2\ NO_{2(g)}$
    (3) $3\ NO_{2(g)} + H_2O_{(l)} \rightarrow 2\ HNO_{3(aq)} + NO_{(g)}$

# Chemical Equation Calculations

CHAPTER **9**

**Section 9.1** *Interpreting a Chemical Equation*

1. Quantities proportional to the coefficients in a balanced chemical equation:
   (1) number of molecules or formula units of reactants and products
   (2) numbers of moles of reactants and products
   (3) volumes of gaseous reactants and products (at the same conditions)

3. General Equation: $A + 2\,B \rightarrow C + 3\,D$

   (a) $5\ \cancel{\text{formula units A}} \times \dfrac{2\ \text{formula units B}}{1\ \cancel{\text{formula unit A}}} = 10$ formula units of B

   (b) $4\ \cancel{\text{moles A}} \times \dfrac{1\ \text{mole C}}{1\ \cancel{\text{mole A}}} = 4$ moles of C

   (c) $2\ \cancel{\text{liters A}} \times \dfrac{3\ \text{liters D}}{1\ \cancel{\text{liter A}}} = 6$ liters D

   (d) $6\ \cancel{\text{molar masses D}} \times \dfrac{2\ \text{molar masses B}}{3\ \cancel{\text{molar masses D}}} = 4$ molar masses of B

5. (a) $2\,KNO_{3(s)} \rightarrow 2\,KNO_{2(s)} + O_{2(g)}$

   MM of $KNO_3$ = 39.1 g K + 14.0 g N + 3(16.0) g O = 101.1 g/mol
   MM of $KNO_2$ = 39.1 g K + 14.0 g N + 2(16.0) g O = 85.1 g/mol
   MM of $O_2$ = 2(16.0) g O = 32.0 g/mol

   2(101.1) g → 2(85.1) g + 1(32.0) g
   202.2 g → 202.2 g

   (b) $2\,Al_{(s)} + 3\,S_{(s)} \rightarrow Al_2S_{3(s)}$

   MM of Al = 27.0 g/mol
   MM of S = 32.1 g/mol
   MM of $Al_2S_3$ = 2(27.0) g Al + 3(32.1) g S = 150.3 g/mol

   2(27.0) g + 3(32.1) g → 150.3 g
   150.3 g → 150.3 g

**Section 9.2** *Mole-Mole Problems*

7. $2\ H_{2(g)} + O_{2(g)} \rightarrow 2\ H_2O_{(g)}$

Moles of $O_2$ that react:

$$0.500\ \cancel{\text{mol } H_2} \times \frac{1 \text{ mol } O_2}{2\ \cancel{\text{mol } H_2}} = 0.250 \text{ mole } O_2 \text{ react}$$

Moles of $H_2O$ produced:

$$0.500\ \cancel{\text{mol } H_2} \times \frac{2 \text{ mol } H_2O}{2\ \cancel{\text{mol } H_2}} = 0.500 \text{ mole } H_2O \text{ produced}$$

9. $2\ Fe_{(s)} + 3\ Cl_{2(g)} \rightarrow 2\ FeCl_{3(s)}$

Moles of $Cl_2$ that react:

$$0.333\ \cancel{\text{mol Fe}} \times \frac{3 \text{ mol } Cl_2}{2\ \cancel{\text{mol Fe}}} = 0.500 \text{ mol } Cl_2 \text{ react}$$

Moles of $FeCl_3$ produced:

$$0.333\ \cancel{\text{mol Fe}} \times \frac{2 \text{ mol } FeCl_3}{2\ \cancel{\text{mol Fe}}} = 0.333 \text{ mol } FeCl_3 \text{ produced}$$

11. $4\ NH_{3(g)} + 5\ O_{2(g)} \rightarrow 4\ NO_{(g)} + 6\ H_2O_{(g)}$

Moles of $NH_3$ that react:

$$1.50\ \cancel{\text{mol NO}} \times \frac{4 \text{ mol } NH_3}{4\ \cancel{\text{mol NO}}} = 1.50 \text{ mol } NH_3 \text{ react}$$

Moles of $H_2O$ produced:

$$1.50\ \cancel{\text{mol NO}} \times \frac{6 \text{ mol } H_2O}{4\ \cancel{\text{mol NO}}} = 2.25 \text{ mol } H_2O \text{ produced}$$

Moles of $O_2$ that react:

$$1.50\ \cancel{\text{mol NO}} \times \frac{5 \text{ mol } O_2}{4\ \cancel{\text{mol NO}}} = 1.88 \text{ mol } O_2 \text{ react}$$

**Section 9.3** *Types of Stoichiometry Problems*

13.

| | Given Quantity | Desired Quantity | Type of Stoichiometry Problem |
|---|---|---|---|
| (a) | mass | mass | mass-mass problem |
| (b) | volume | volume | volume-volume problem |
| (c) | mass | volume | mass-volume problem |

**Section 9.4** *Mass-Mass Stoichiometry Problems*

15. $2\ Zn_{(s)} + O_{2(g)} \rightarrow 2\ ZnO_{(s)}$

MM of Zn = 65.4 g/mol
MM of ZnO = 81.4 g/mol

$$2.36\ \cancel{g\ Zn} \times \frac{1\ \cancel{mol\ Zn}}{65.4\ \cancel{g\ Zn}} \times \frac{2\ \cancel{mol\ ZnO}}{2\ \cancel{mol\ Zn}} \times \frac{81.4\ g\ ZnO}{1\ \cancel{mol\ ZnO}} = 2.94\ g\ ZnO$$

17. $2\ Bi_{(s)} + 3\ Cl_{2(g)} \rightarrow 2\ BiCl_{3(s)}$

MM of Bi = 209.0 g/mol
MM of $BiCl_3$ = 315.5 g/mol

$$3.45\ \cancel{g\ Bi} \times \frac{1\ \cancel{mol\ Bi}}{209.0\ \cancel{g\ Bi}} \times \frac{2\ \cancel{mol\ BiCl_3}}{2\ \cancel{mol\ Bi}} \times \frac{315.5\ g\ BiCl_3}{1\ \cancel{mol\ BiCl_3}} = 5.21\ g\ BiCl_3$$

19. $Cu_{(s)} + 2\ AgNO_{3(aq)} \rightarrow Cu(NO_3)_{2(aq)} + 2\ Ag_{(s)}$

MM of Cu = 63.5 g/mol
MM of Ag = 107.9 g/mol

$$0.615\ \cancel{g\ Cu} \times \frac{1\ \cancel{mol\ Cu}}{63.5\ \cancel{g\ Cu}} \times \frac{2\ \cancel{mol\ Ag}}{1\ \cancel{mol\ Cu}} \times \frac{107.9\ g\ Ag}{1\ \cancel{mol\ Ag}} = 2.09\ g\ Ag$$

21. $2\ Co_{(s)} + 3\ HgCl_{2(aq)} \rightarrow 2\ CoCl_{3(aq)} + 3\ Hg_{(l)}$

MM of Co = 58.9 g/mol
MM of Hg = 200.6 g/mol

$$1.25\ \cancel{g\ Co} \times \frac{1\ \cancel{mol\ Co}}{58.9\ \cancel{g\ Co}} \times \frac{3\ \cancel{mol\ Hg}}{2\ \cancel{mol\ Co}} \times \frac{200.6\ g\ Hg}{1\ \cancel{mol\ Hg}} = 6.39\ g\ Hg$$

23. $2\ Na_3PO_{4(aq)} + 3\ Ca(OH)_{2(aq)} \rightarrow Ca_3(PO_4)_{2(s)} + 6\ NaOH_{(aq)}$

MM of $Na_3PO_4$ = 164.0 g/mol
MM of $Ca_3(PO_4)_2$ = 310.3 g/mol

$$1.78\ \cancel{g\ Na_3PO_4} \times \frac{1\ \cancel{mol\ Na_3PO_4}}{164.0\ \cancel{g\ Na_3PO_4}} \times \frac{1\ \cancel{mol\ Ca_3(PO_4)_2}}{2\ \cancel{mol\ Na_3PO_4}} \times \frac{310.3\ g\ Ca_3(PO_4)_2}{1\ \cancel{mol\ Ca_3(PO_4)_2}}$$
$$= 1.68\ g\ Ca_3(PO_4)_2$$

## Section 9.5 *Mass-Volume Stoichiometry Problems*

25. $Fe_2(CO_3)_{3(s)} \rightarrow Fe_2O_{3(s)} + 3\ CO_{2(g)}$

MM of $Fe_2(CO_3)_3$ = 291.6 g/mol

$$1.59\ \cancel{g\ Fe_2(CO_3)_3} \times \frac{1\ \cancel{mol\ Fe_2(CO_3)_3}}{291.6\ \cancel{g\ Fe_2(CO_3)_3}} \times \frac{3\ \cancel{mol\ CO_2}}{1\ \cancel{mol\ Fe_2(CO_3)_3}} \times \frac{22.4\ \cancel{L\ CO_2}}{1\ \cancel{mol\ CO_2}} \times \frac{1000\ mL\ CO_2}{1\ \cancel{L\ CO_2}} = 366\ mL\ CO_2$$

27. $2\ LiHCO_{3(s)} \rightarrow Li_2CO_{3(s)} + H_2O_{(l)} + CO_{2(g)}$

MM of $LiHCO_3$ = 67.9 g/mol

$$1.59\ \cancel{g\ LiHCO_3} \times \frac{1\ \cancel{mol\ LiHCO_3}}{67.9\ \cancel{g\ LiHCO_3}} \times \frac{1\ \cancel{mol\ CO_2}}{2\ \cancel{mol\ LiHCO_3}} \times \frac{22.4\ \cancel{L\ CO_2}}{1\ \cancel{mol\ CO_2}} \times \frac{1000\ mL\ CO_2}{1\ \cancel{L\ CO_2}} = 262\ mL\ CO_2$$

29. $Mg_{(s)} + H_2SO_{4(aq)} \rightarrow MgSO_{4(aq)} + H_{2(g)}$

MM of Mg = 24.3 g/mol

$$225\ \cancel{mL\ H_2} \times \frac{1\ \cancel{L\ H_2}}{1000\ \cancel{mL\ H_2}} \times \frac{1\ \cancel{mol\ H_2}}{22.4\ \cancel{L\ H_2}} \times \frac{1\ \cancel{mol\ Mg}}{1\ \cancel{mol\ H_2}} \times \frac{24.3\ g\ Mg}{1\ \cancel{mol\ Mg}} = 0.244\ g\ Mg$$

31. $2\ H_2O_{2(l)} \rightarrow 2\ H_2O_{(l)} + O_{2(g)}$

MM of $H_2O_2$ = 34.0 g/mol

$$55.0\ \cancel{mL\ O_2} \times \frac{1\ \cancel{L\ O_2}}{1000\ \cancel{mL\ O_2}} \times \frac{1\ \cancel{mol\ O_2}}{22.4\ \cancel{L\ O_2}} \times \frac{2\ \cancel{mol\ H_2O_2}}{1\ \cancel{mol\ O_2}} \times \frac{34.0\ g\ H_2O_2}{1\ \cancel{mol\ H_2O_2}} = 0.167\ g\ H_2O_2$$

## Section 9.6 *Volume-Volume Stoichiometry Problems*

33. $2\ CO_{(g)} + O_{2(g)} \rightarrow 2\ CO_{2(g)}$

$$2.00\ \cancel{L\ CO} \times \frac{1\ L\ O_2}{2\ \cancel{L\ CO}} = 1.00\ L\ O_2$$

35. $3\ H_{2(g)} + N_{2(g)} \rightarrow 2\ NH_{3(g)}$

(a) $38.5\ \cancel{mL\ NH_3} \times \dfrac{1\ mL\ N_2}{2\ \cancel{mL\ NH_3}} = 19.3\ mL\ N_2$

(b) $38.5\ \cancel{mL\ NH_3} \times \dfrac{3\ mL\ H_2}{2\ \cancel{mL\ NH_3}} = 57.8\ mL\ H_2$

37. $2\ SO_{2(g)} + O_{2(g)} \rightarrow 2\ SO_{3(g)}$

$25.0\ \cancel{L\ O_2} \times \dfrac{2\ L\ SO_3}{1\ \cancel{L\ O_2}} = 50.0\ L\ SO_3$

**Section 9.7** *Percent Yield*

39. Actual yield: 2.01 g $PbI_{2(s)}$
Theoretical yield: 2.16 g $PbI_{2(s)}$

Percent yield: $\dfrac{2.01\ \cancel{g}}{2.16\ \cancel{g}} \times 100 = 93.1\%$

41. Actual yield: 1.29 g $NaNO_2$
Theoretical yield: 1.22 g $NaNO_2$

Percent yield: $\dfrac{1.29\ \cancel{g}}{1.22\ \cancel{g}} \times 100 = 106\%$

**Section 9.8** *Experimental Accuracy and Precision*

43. (a) Error: $\left(\dfrac{19.5\% + 20.4\% + 19.1\%}{3}\right) - 21.7\% = -2.0\%$

(b) Range: $20.4\% - 19.1\% = 1.3\%$

45. (a) Accuracy: $\left(\dfrac{93.5\% + 97.0\% + 96.9\%}{3}\right) - 95.0\% = 0.8\%$

(b) Precision: $97.0\% - 93.5\% = 3.5\%$

## General Exercises

47. $(NH_4)_2Cr_2O_{7(s)} \rightarrow Cr_2O_{3(s)} + N_{2(g)} + 4\ H_2O_{(l)}$

MM of $(NH_4)_2Cr_2O_7$ = 252.0 g/mol
MM of $Cr_2O_3$ = 152.0 g/mol

(a) $$1.54\ \cancel{g\ (NH_4)_2Cr_2O_7} \times \frac{1\ \cancel{mol\ (NH_4)_2Cr_2O_7}}{252.0\ \cancel{g\ (NH_4)_2Cr_2O_7}} \times \frac{1\ \cancel{mol\ Cr_2O_3}}{1\ \cancel{mol\ (NH_4)_2Cr_2O_7}} \times \frac{152.0\ g\ Cr_2O_3}{1\ \cancel{mol\ Cr_2O_3}} = 0.929\ g\ Cr_2O_3$$

(b) $$1.54\ \cancel{g\ (NH_4)_2Cr_2O_7} \times \frac{1\ \cancel{mol\ (NH_4)_2Cr_2O_7}}{252.0\ \cancel{g\ (NH_4)_2Cr_2O_7}} \times \frac{1\ \cancel{mol\ N_2}}{1\ \cancel{mol\ (NH_4)_2Cr_2O_7}} \times \frac{22.4\ \cancel{L\ N_2}}{1\ \cancel{mol\ N_2}} \times \frac{1000\ mL\ N_2}{1\ \cancel{L\ N_2}} = 137\ mL\ N_2$$

49. $Sb_2S_{3(s)} + 6\ HCl_{(aq)} \rightarrow 2\ SbCl_{3(aq)} + 3\ H_2S_{(g)}$

MM of $Sb_2S_3$ = 339.9 g/mol

$$3.00\ \cancel{g\ Sb_2S_3} \times \frac{1\ \cancel{mol\ Sb_2S_3}}{339.9\ \cancel{g\ Sb_2S_3}} \times \frac{3\ \cancel{mol\ H_2S}}{1\ \cancel{mol\ Sb_2S_3}} \times \frac{22.4\ L\ H_2S}{1\ \cancel{mol\ H_2S}} = 0.593\ L\ H_2S$$

51. $C_3H_{8(g)} + 5\ O_{2(g)} \rightarrow 3\ CO_{2(g)} + 4\ H_2O_{(g)}$

MM of $C_3H_8$ = 44.0 g/mol
MM of $H_2O$ = 18.0 g/mol

(a) $$10.0\ \cancel{g\ C_3H_8} \times \frac{1\ \cancel{mol\ C_3H_8}}{44.0\ \cancel{g\ C_3H_8}} \times \frac{4\ \cancel{mol\ H_2O}}{1\ \cancel{mol\ C_3H_8}} \times \frac{18.0\ g\ H_2O}{1\ \cancel{mol\ H_2O}} = 16.4\ g\ H_2O$$

(b) $$10.0\ \cancel{g\ C_3H_8} \times \frac{1\ \cancel{mol\ C_3H_8}}{44.0\ \cancel{g\ C_3H_8}} \times \frac{3\ \cancel{mol\ CO_2}}{1\ \cancel{mol\ C_3H_8}} \times \frac{22.4\ \cancel{L\ CO_2}}{1\ \cancel{mol\ CO_2}} \times \frac{1000\ mL\ CO_2}{1\ \cancel{L\ CO_2}} = 15{,}300\ mL\ CO_2$$

# Modern Atomic Theory

CHAPTER 10

## Section 10.1 *Bohr Model of the Atom*

1. (a) True
   (b) False (Electrons in the Bohr Model are at fixed energy states and do not lose energy as they orbit the nucleus.)

3. Of the given lines from the emission spectrum of hydrogen, violet is the most energetic.

5.

| | Number of Electrons | Energy Level Change | Number of Quanta |
|---|---|---|---|
| (a) | 1 $e^-$ | 3 to 1 | 1 quantum |
| (b) | 1 $e^-$ | 3 to 2 | 1 quantum |
| (c) | 100 $e^-$ | 3 to 2 | 100 quanta |
| (d) | 100 $e^-$ | 4 to 2 | 100 quanta |
| (e) | 500 $e^-$ | 5 to 2 | 500 quanta |
| (f) | 500 $e^-$ | 5 to 3 | 500 quanta |

7. The Bohr atom successfully explained the spectral lines in the emission spectrum of hydrogen.

## Section 10.2 *Quantum Theory*

9.

| | Object | Spectrum |
|---|---|---|
| (a) | Rainbow | continuous |
| (b) | emission line spectrum | quantized |

11.

| | Instrument | Measurement |
|---|---|---|
| (a) | 10-mL graduated cylinder | continuous |
| (b) | 10-mL volumetric pipet | quantized |

13. The photons of infrared light are not sufficiently energetic, whereas the photons of ultraviolet light have sufficient energy to dislodge electrons.

15. An orbit is the path traveled by an electron about the nucleus of an atom, according to the Bohr model.

    An orbital is a region about the nucleus in which there is a high probability of finding an electron with a given energy, according to the quantum mechanical model.

17. The uncertainty principle, formulated by Heisenberg, states that it is impossible to precisely measure both the location and momentum of a small particle simultaneously.

19.

| | Principal Energy Levels | Most Energetic |
|---|---|---|
| (a) | $n = 1$ or $n = 2$ | $n = 2$ |
| (b) | $n = 1$ or $n = 3$ | $n = 3$ |
| (c) | $n = 2$ or $n = 3$ | $n = 3$ |
| (d) | $n = 3$ or $n = 5$ | $n = 5$ |

21.

| | Identical |
|---|---|
| (b) | $p$ and $l = 1$ |
| (c) | $d$ and $l = 2$ |

23.* (a) (b)

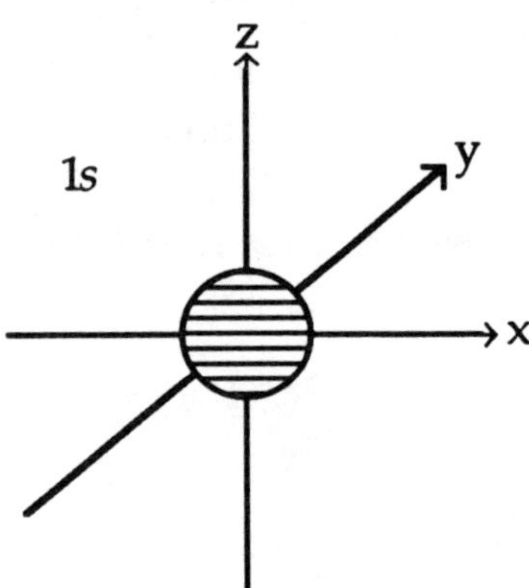


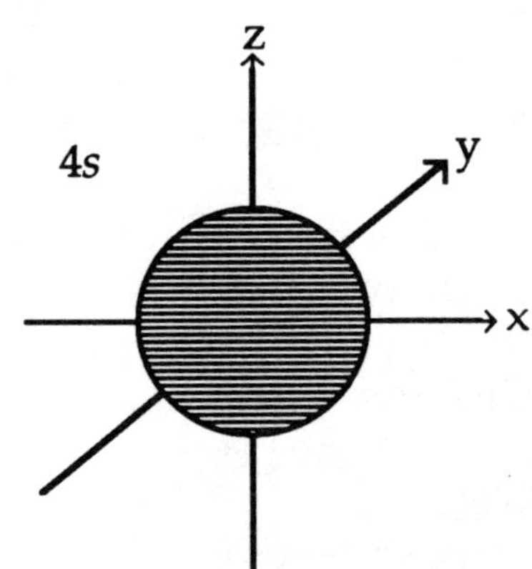


* continued on next page

23. (c)

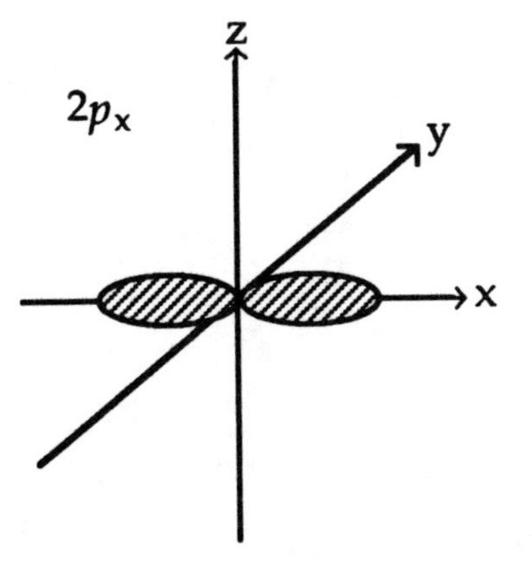


(d)

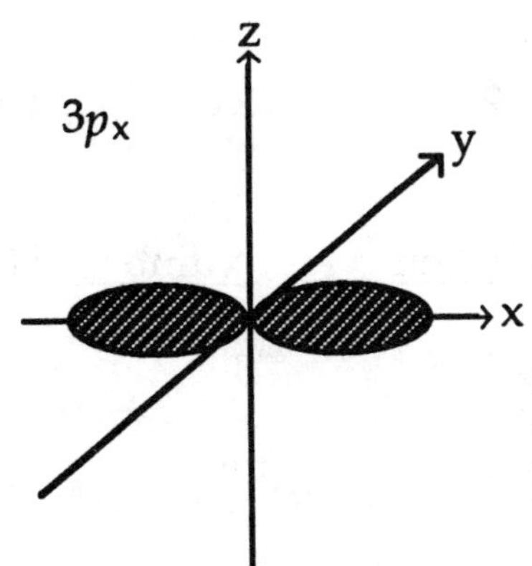


(e)

$3p_y$

z

y

x

(f)

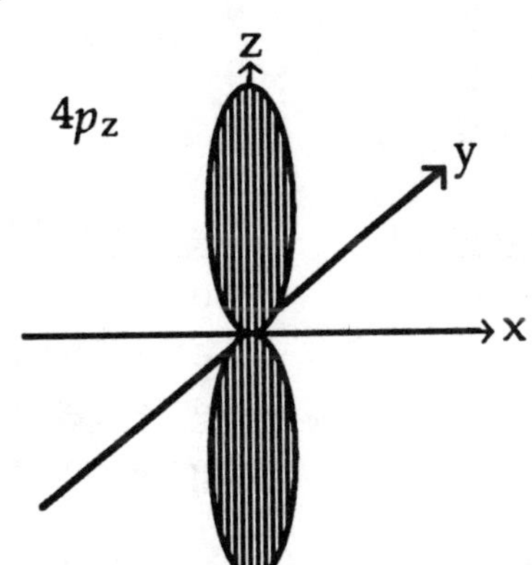


25.

| | Orbitals | Higher Energy | | Orbitals | Higher Energy |
|---|---|---|---|---|---|
| (a) | 2*s* or 3*s* | 3*s* | (b) | $2p_x$ or $3p_x$ | $3p_x$ |
| (c) | $2p_x$ or $2p_y$ | both are equal | (d) | $4p_y$ or $4p_z$ | both are equal |

27.

| | Description | Orbital |
|---|---|---|
| (a) | spherical orbital in the fifth shell | 5*s* |
| (b) | dumbbell-shaped orbital in the fourth shell | 4*p* |

## Section 10.4 *Distribution of Electrons by Orbital*

29.

| | Orbital | Max. # of Electrons |
|---|---|---|
| (a) | an *s* orbital | 2 $e^-$ |
| (b) | a *p* orbital | 2 $e^-$ |
| (c) | a *d* orbital | 2 $e^-$ |
| (d) | an *f* orbital | 2 $e^-$ |

| 31. | Orbital | Max. # of Electrons |
|---|---|---|
| (a) | $l = 0$ | 2 $e^-$ |
| (b) | $l = 1$ | 2 $e^-$ |
| (c) | $l = 2$ | 2 $e^-$ |
| (d) | $l = 3$ | 2 $e^-$ |

| 33. | Subshell | Number of Orbitals |
|---|---|---|
| (a) | 3*s* | 1 |
| (b) | 4*p* | 3 |
| (c) | 3*d* | 5 |
| (d) | 4*f* | 7 |

| 35. | Shell | Number of Subshells |
|---|---|---|
| (a) | $n = 1$ | 1 |
| (b) | $n = 2$ | 2 |
| (c) | $n = 3$ | 3 |
| (d) | $n = 4$ | 4 |

| 37. | Subshell | Max. # of Electrons |
|---|---|---|
| (a) | 4*s* | 2 $e^-$ |
| (b) | 4*p* | 2 $e^-$ + 2 $e^-$ + 2 $e^-$ = 6 $e^-$ |
| (c) | 4*d* | 2 $e^-$ + 2 $e^-$ + 2 $e^-$ + 2 $e^-$ + 2 $e^-$ = 10 $e^-$ |
| (d) | 4*f* | 2 $e^-$ + 2 $e^-$ + 2 $e^-$ + 2 $e^-$ + 2 $e^-$ + 2 $e^-$ + 2 $e^-$ = 14 $e^-$ |

| 39. | Shell | Max. # of Electrons |
|---|---|---|
| (a) | $n = 1$ | 2 $e^-$ |
| (b) | $n = 2$ | 2 $e^-$ + 6 $e^-$ = 8 $e^-$ |
| (c) | $n = 3$ | 2 $e^-$ + 6 $e^-$ + 10 $e^-$ = 18 $e^-$ |
| (d) | $n = 4$ | 2 $e^-$ + 6 $e^-$ + 10 $e^-$ + 14 $e^-$ = 32 $e^-$ |

## **Section 10.5** *The Four Quantum Numbers*

41. $n = 1, 2, 3, 4, 5, 6, 7, \ldots$

| 43. | Principal Quantum Number | 2nd Quantum Number |
|---|---|---|
| (a) | $n = 1$ | $l = 0$ |
| (b) | $n = 2$ | $l = 0, 1$ |
| (c) | $n = 3$ | $l = 0, 1, 2$ |
| (d) | $n = 4$ | $l = 0, 1, 2, 3$ |

| 45. | 2nd Quantum Number | 3rd Quantum Number |
|---|---|---|
| (a) | $l = 0$ | $m = 0$ |
| (b) | $l = 1$ | $m = +1, 0, -1$ |
| (c) | $l = 2$ | $m = +2, +1, 0, -1, -2$ |
| (d) | $l = 3$ | $m = +3, +2, +1, 0, -1, -2, -3$ |

47. Orbital / 3rd Quantum Number

| | Orbital | 3rd Quantum Number |
|---|---|---|
| (a) | 4*s* | $m = 0$ |
| (b) | 4*p* | $m = +1, 0, -1$ |
| (c) | 4*d* | $m = +2, +1, 0, -1, -2$ |
| (d) | 4*f* | $m = +3, +2, +1, 0, -1, -2, -3$ |

49.

| | 3rd Quantum Number | 4th Quantum Number |
|---|---|---|
| (a) | $m = 0$ | $s = +\frac{1}{2}, -\frac{1}{2}$ |
| (b) | $m = +1$ | $s = +\frac{1}{2}, -\frac{1}{2}$ |
| (c) | $m = +2$ | $s = +\frac{1}{2}, -\frac{1}{2}$ |
| (d) | $m = -3$ | $s = +\frac{1}{2}, -\frac{1}{2}$ |

51. Orbital energy and size are specified by the principal quantum number. (Note: As the principal quantum number increases, so do the size and energy of the orbitals.)

53. The orientation of an orbital is specified by the third quantum number.

**Section 10.6** ***Orbital Energy Diagrams***

55.

| | Element | Electron Configuration |
|---|---|---|
| (a) | C | $1s^2\,2s^2\,2p^2$ |
| (b) | P | $1s^2\,2s^2\,2p^6\,3s^2\,3p^3$ |
| (c) | K | $1s^2\,2s^2\,2p^6\,3s^2\,3p^6\,4s^1$ |
| (d) | Co | $1s^2\,2s^2\,2p^6\,3s^2\,3p^6\,4s^2\,3d^7$ |

57. Orbital Diagram

(a)

| 1*s* | 2*s* | 2*p* |
|---|---|---|
| [↑↓] | [↑↓] | [↑] [↑] [ ] |

(b)

| 1*s* | 2*s* | 2*p* | 3*s* | 3*p* |
|---|---|---|---|---|
| [↑↓] | [↑↓] | [↑↓] [↑↓] [↑↓] | [↑↓] | [↑] [↑] [↑] |

(c)

| 1*s* | 2*s* | 2*p* | 3*s* | 3*p* | 4*s* |
|---|---|---|---|---|---|
| [↑↓] | [↑↓] | [↑↓] [↑↓] [↑↓] | [↑↓] | [↑↓] [↑↓] [↑↓] | [↑] |

(d)

| 1*s* | 2*s* | 2*p* | 3*s* | 3*p* | 4*s* | 3*d* |
|---|---|---|---|---|---|---|
| [↑↓] | [↑↓] | [↑↓] [↑↓] [↑↓] | [↑↓] | [↑↓] [↑↓] [↑↓] | [↑↓] | [↑↓] [↑↓] [↑] [↑] [↑] |

59.

| | 1*s* | 2*s* | 2*p* | 3*s* | 3*p* |
|---|---|---|---|---|---|
| S | ↑↓ | ↑↓ | ↑↓ ↑↓ ↑↓ | ↑↓ | ↑↓ ↑ ↑ |

Sulfur has two unpaired electrons in the 3*p* subshell.

61. Orbital Energy Diagram

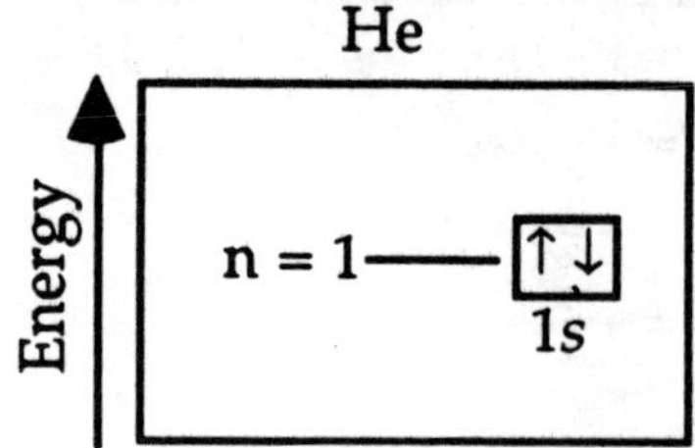


63.* Orbital Energy Diagrams

(a)

Li

Energy

n = 2 — ↑ 2*s*

n = 1 — ↑↓ 1*s*

(b)

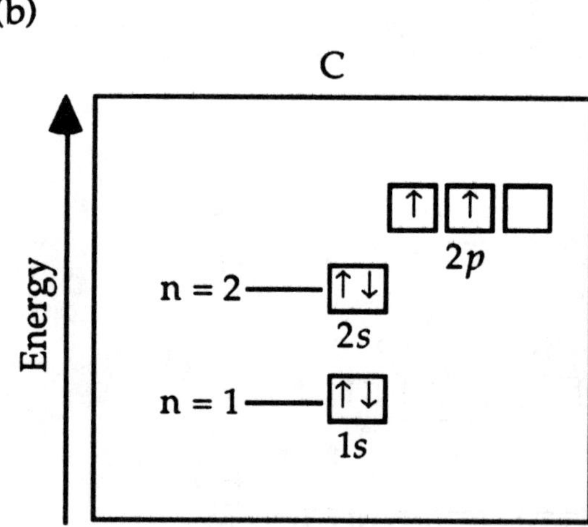


* continued on next page

63. (c)

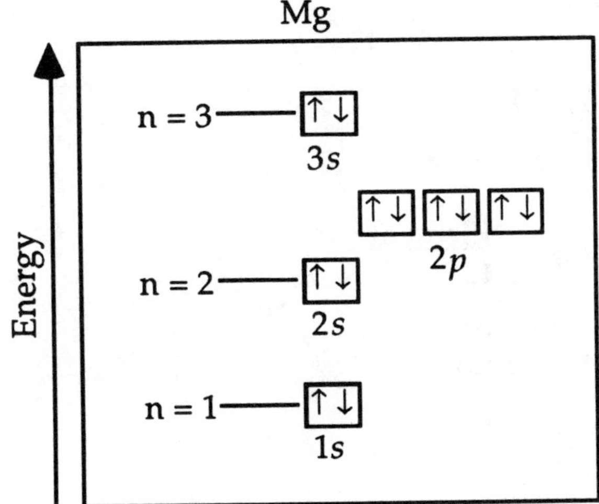


(d)

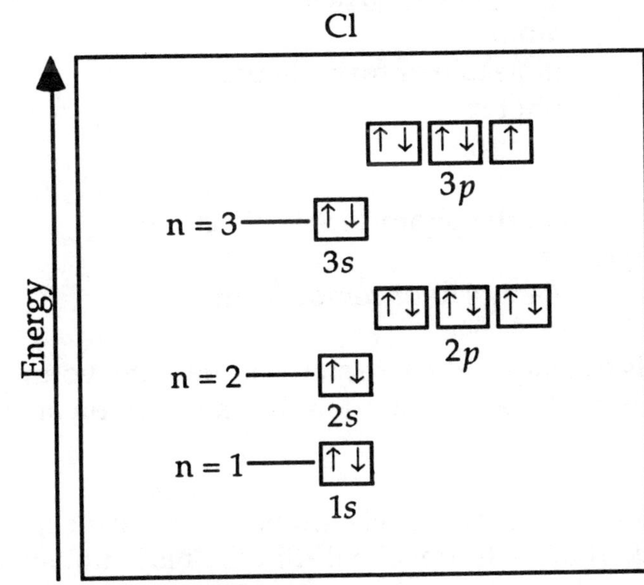


**Section 10.7** *Quantum Notation: (n, l, m, s)*

65.

| | Element | Quantum Numbers for Highest-Energy Electron |
|---|---|---|
| (a) | boron | $(2, 1, +1, +\frac{1}{2})$ |
| (b) | neon | $(2, 1, -1, -\frac{1}{2})$ |
| (c) | magnesium | $(3, 0, 0, -\frac{1}{2})$ |
| (d) | oxygen | $(2, 1, +1, -\frac{1}{2})$ |

67.

| | Element | Quantum Numbers for Highest-Energy Electron |
|---|---|---|
| (a) | Na | $(3, 0, 0, +\frac{1}{2})$ |
| (b) | F | $(2, 1, 0, -\frac{1}{2})$ |
| (c) | Sc | $(3, 2, +2, +\frac{1}{2})$ |
| (d) | Kr | $(4, 1, -1, -\frac{1}{2})$ |

69.

| | Quantum Notation | Element |
|---|---|---|
| (a) | $(2, 0, 0, +\frac{1}{2})$ | Li |
| (b) | $(3, 1, +1, -\frac{1}{2})$ | S |
| (c) | $(4, 2, 0, +\frac{1}{2})$ | Nb |
| (d) | $(5, 1, 0, -\frac{1}{2})$ | I |

## General Exercises

71.

| | Matter/Energy | Quantized Particle |
|---|---|---|
| (a) | element | atom |
| (b) | compound | molecule or formula unit |
| (c) | light energy | photon |
| (d) | electrical energy | electron |

73. Some practical applications of the photoelectric effect are:
(1) automatic door openers
(2) television, stereo, and garage door remote controls

75. Since 2*s* and 2*p* orbitals overlap and share some of the same volume, the electrons in 2*s* and 2*p* orbitals can be found in the same region about the nucleus.

77. If chromium promotes one of its two 4*s* electrons into a vacant 3*d* orbital, the resulting 4*s* and 3*d* orbital sets are all half-filled. This is unusually stable.

79. | Element | Orbital Diagram of Highest-Energy Electrons
--- | --- | ---
(a) | Ti | 3d: [↑][↑][ ][ ][ ]
(b) | Zn | 3d: [↑↓][↑↓][↑↓][↑↓][↑↓]
(c) | Ni | 3d: [↑↓][↑↓][↑↓][↑][↑]
(d) | Ba | 6s: [↑↓]

Of the previous four elements, only Ti and Ni have unpaired electrons. Therefore, only Ti and Ni are paramagnetic.

CHAPTER

# 11 Chemical Bonding

## Section 11.1 *Valence Electrons and Chemical Bonds*

1. When a metal atom loses valence electrons to become a positively-charged cation and a nonmetal atom gains valence electrons to become a negatively-charged anion the resulting attraction between the two ions is an ionic bond.

3.

| Element | Before Reaction | After Reaction |
|---|---|---|
| Mg | 2 valence $e^-$ | 0 valence $e^-$ |
| S | 6 valence $e^-$ | 8 valence $e^-$ |

5.

| | Compound | Bond | | Compound | Bond |
|---|---|---|---|---|---|
| (a) | $H_2O$ | covalent | (b) | NaCl | ionic |
| (c) | $CH_4$ | covalent | (d) | CdO | ionic |

7.

| | Substance | Representative Particle |
|---|---|---|
| (a) | $C_2H_5OH$ | molecule |
| (b) | $Au_2S_3$ | formula unit |
| (c) | $PH_3$ | molecule |
| (d) | $N_2O_4$ | molecule |
| (e) | $OF_2$ | molecule |
| (f) | $K_2CO_3$ | formula unit |

9.

| | Substance | Representative Particle |
|---|---|---|
| (a) | Ne | atom |
| (b) | $Cl_2$ | molecule |
| (c) | $C_{17}H_{21}NO_4$ | molecule |
| (d) | Fe | atom |
| (e) | $P_4$ | molecule |
| (f) | $PuBr_3$ | formula unit |

**Section 11.2** *The Ionic Bond*

11.

| | Ion | Charge | | Ion | Charge |
|---|---|---|---|---|---|
| (a) | Na ion | 1+ | (b) | Cs ion | 1+ |
| (c) | Sn ion | 4+ | (d) | Pb ion | 4+ |

13.

| | Ion | Charge | | Ion | Charge |
|---|---|---|---|---|---|
| (a) | F ion | 1– | (b) | Br ion | 1– |
| (c) | S ion | 2– | (d) | N ion | 3– |

15.

| | Ion | Electron Configuration |
|---|---|---|
| (a) | $Li^+$ | $1s^2$ |
| (b) | $K^+$ | $1s^2\ 2s^2\ 2p^6\ 3s^2\ 3p^6$ |
| (c) | $Ca^{2+}$ | $1s^2\ 2s^2\ 2p^6\ 3s^2\ 3p^6$ |
| (d) | $Ra^{2+}$ | $1s^2\ 2s^2\ 2p^6\ 3s^2\ 3p^6\ 4s^2\ 3d^{10}\ 4p^6\ 5s^2\ 4d^{10}\ 5p^6\ 6s^2 4f^{14}\ 5d^{10}\ 6p^6$ |

17.

| | Ion | Electron Configuration |
|---|---|---|
| (a) | $Cl^-$ | $1s^2\ 2s^2\ 2p^6\ 3s^2\ 3p^6$ |
| (b) | $I^-$ | $1s^2\ 2s^2\ 2p^6\ 3s^2\ 3p^6\ 4s^2\ 3d^{10}\ 4p^6\ 5s^2\ 4d^{10}\ 5p^6$ |
| (c) | $S^{2-}$ | $1s^2\ 2s^2\ 2p^6\ 3s^2\ 3p^6$ |
| (d) | $P^{3-}$ | $1s^2\ 2s^2\ 2p^6\ 3s^2\ 3p^6$ |

19.

| | Ion | Isoelectronic With: |
|---|---|---|
| (a) | $Li^+$ | He |
| (b) | $K^+$ | Ar |
| (c) | $Ca^{2+}$ | Ar |
| (d) | $Ra^{2+}$ | Rn |

21.

| | Ion | Isoelectronic With: |
|---|---|---|
| (a) | $Sc^{3+}$ | Ar |
| (b) | $Y^{3+}$ | Kr |
| (c) | $Ti^{4+}$ | Ar |
| (d) | $Zr^{4+}$ | Kr |

23. (a) Li atomic radius > Li ion radius
(b) Mg atomic radius > Mg ion radius
(c) F atomic radius < F ion radius
(d) O atomic radius < O ion radius

25. All of the statements are false. The corrected statements are:

(a) An ionic bond can be formed by the transfer of valence electrons from the *metal* atom to the *nonmetal* atom.

(b) The ionic radius of a metal atom is *less* than its atomic radius.

(c) The ionic radius of a nonmetal atom is *greater* than its atomic radius.

(d) The properties of a compound are *not similar to* the properties of the constituent elements.

(e) Energy is released when an ionic bond is *formed from* ions.

**Section 11.3** *The Covalent Bond*

27. (a) The sum of the H and I atomic radii is greater than the bond length in H—I.

(b) The sum of the N and O atomic radii is greater than the bond length in N—O.

29. All of the statements are false. The corrected statements are:

(a) A covalent bond is formed by *sharing a pair of electrons between two nonmetal atoms.*

(b) The bonding electrons are *delocalized over both atoms* in the covalent bond.

(c) The bond length in a covalent bond is *less than* the sum of the two atomic radii.

(d) The properties of a compound are *not similar to* the properties of the constituent elements.

(e) An amount of energy equal to the bond energy *is required to break* a covalent bond.

**Section 11.4** *Electron Dot Formulas of Molecules*

31.*

| | Molecule | Valence Electrons | Electron Dot | Structural |
|---|---|---|---|---|
| (a) | $H_2$ | $1 + 1 = 2\ e^-$ | H:H | H—H |
| (b) | $F_2$ | $7 + 7 = 14\ e^-$ | :F̤̈:F̤̈: | F—F |
| (c) | HBr | $1 + 7 = 8\ e^-$ | H:B̤̈r: | H—Br |
| (d) | $NH_3$ | $5 + 3(1) = 8\ e–$ | H<br>H:N̤:H | H<br>\|<br>H—N—H |

* continued on next page

31.

| | Molecule | Valence Electrons | Electron Dot | Structural |
|---|---|---|---|---|
| (e) | $CCl_4$ | $4 + 4(7) = 32\ e^-$ | :Cl: / :Cl:C:Cl: / :Cl: | Cl–C(–Cl)(–Cl)–Cl |
| (f) | $OF_2$ | $6 + 2(7) = 20\ e^-$ | :F:O:F: | F—O—F |

33.

| | Molecule | Valence Electrons | Electron Dot | Structural |
|---|---|---|---|---|
| (a) | $HONO_2$ | $1 + 3(6) + 5 = 24\ e^-$ | H:O:N::O / :O: | H—O—N=O (with N—O) |
| (b) | $SO_2$ | $3(6) = 18\ e^-$ | :O:S::O | O—S=O |
| (c) | $C_2H_4$ | $2(4) + 4(1) = 12\ e^-$ | H:C::C:H, H H | H—C=C—H, H H |
| (d) | $C_2H_2$ | $2(4) + 2(1) = 10\ e^-$ | H:C:::C:H | H—C≡C—H |
| (e) | HOCl | $1 + 6 + 7 = 14\ e^-$ | H:O:Cl: | H—O—Cl |
| (f) | HONO | $1 + 2(6) + 5 = 18\ e^-$ | H:O:N::O | H—O—N=O |

**Section 11.5** *Polar Covalent Bonds*

35. Electronegativity down a group in the periodic table generally decreases.

37. Nonmetals are more electronegative than metals.

39.

| | More Electronegative | | More Electronegative |
|---|---|---|---|
| (a) | **Cl** > Br | (b) | **O** > S |
| (c) | Al < **B** | (d) | **Se** > As |
| (e) | N < **F** | (f) | **P** > Sb |

(Note: The more electronegative element is in bold.)

41.

| | Bond | Polarity | | Bond | Polarity |
|---|---|---|---|---|---|
| (a) | H—Br | 2.8 – 2.1 = 0.7 | (b) | H—F | 4.0 – 2.1 = 1.9 |
| (c) | I—Cl | 3.0 – 2.5 = 0.5 | (d) | Br—F | 4.0 – 2.8 = 1.2 |
| (e) | C—O | 3.5 – 2.5 = 1.0 | | | |

43. Polar Bonds Using Delta Notation

(a) $\delta^+$ H—S $\delta^-$
(b) $\delta^-$ O—S $\delta^+$
(c) $\delta^-$ Cl—B $\delta^+$
(d) $\delta^+$ S—Cl $\delta^-$
(e) $\delta^+$ N—F $\delta^-$

(Note: $\delta^-$ indicates the more electronegative atom and $\delta^+$ indicates the more electropositive atom.)

**Section 11.6** *Nonpolar Covalent Bonds*

45.

| | Bond | Polarity | Classification |
|---|---|---|---|
| (a) | Cl—Cl | 3.0 – 3.0 = 0 | nonpolar |
| (b) | Cl—N | 3.0 – 3.0 = 0 | nonpolar |
| (c) | N—H | 3.0 – 2.1 = 0.9 | polar |
| (d) | H—P | 2.1 – 2.1 = 0 | nonpolar |
| (e) | P—Te | 2.1 – 2.1 = 0 | nonpolar |

Thus, (a), (b), (d), and (e) are nonpolar.

47. Naturally Occurring Nonpolar Diatomic Molecules

(a) hydrogen
(b) nitrogen
(c) chlorine
(f) fluorine
(g) iodine
(i) oxygen
(j) bromine

The element (d) phosphorus occurs naturally as $P_4$, while (e) helium and (h) selenium are monoatomic.

**Section 11.7** *Coordinate Covalent Bonds*

(Note: Coordinate covalent bonds are indicated by a dash, —.)

49.

| Formula | Electron Dot Without O | Formula | Electron Dot With O |
|---|---|---|---|
| HOCl | $\mathrm{H{:}\ddot{\underset{\cdot\cdot}{O}}{:}\ddot{\underset{\cdot\cdot}{Cl}}{:}}$ | HOClO | $\mathrm{H{:}\ddot{\underset{\cdot\cdot}{O}}{:}\ddot{\underset{\cdot\cdot}{Cl}}{-}\ddot{\underset{\cdot\cdot}{O}}{:}}$ |

51.

| Formula | Electron Dot Without H | Formula | Electron Dot With H |
|---|---|---|---|
| $NH_3$ | ••<br>H:N:H<br>••<br>H | $NH_4^+$ | [ H<br>\|<br>H:N:H<br>••<br>H ]$^+$ |

53.

| Formula | Electron Dot Without O | Formula | Electron Dot With O |
|---|---|---|---|
| $NO_2^-$ | [ :Ö:N̈::Ö ]$^-$ | $NO_3^-$ | [ :Ö:<br>\|<br>:Ö:N::Ö ]$^-$ |

55.

| Formula | Electron Dot Without H | Formula | Electron Dot With H |
|---|---|---|---|
| $SO_4^{2-}$ | [ :Ö:<br>:Ö:S̈:Ö:<br>:Ö: ]$^{2-}$ | $HSO_4^-$ | [ :Ö:<br>:Ö:S̈:Ö–H<br>:Ö: ]$^-$ |

**Section 11.8** *Polyatomic Ions*

57.

| | Ion | Valence Electrons | Electron Dot | Structural |
|---|---|---|---|---|
| (a) | **$OH^-$** | $6 + 1 + 1 = 8\ e^-$ | [ :Ö:H ]$^-$ | [ O–H ]$^-$ |
| (b) | **$IO^-$** | $7 + 6 + 1 = 14\ e^-$ | [ :Ï:Ö: ]$^-$ | [ I–O ]$^-$ |
| (c) | **$ClO_2^-$** | $7 + 2(6) + 1 = 20\ e^-$ | [ :Ö:C̈l:Ö: ]$^-$ | [ O–Cl–O ]$^-$ |
| (d) | **$PH_4^+$** | $5 + 4(1) - 1 = 8\ e^-$ | [ H<br>H:P:H<br>H ]$^+$ | [ H<br>\|<br>H–P–H<br>\|<br>H ]$^+$ |
| (e) | **$IO_4^-$** | $7 + 4(6) + 1 = 32\ e^-$ | [ :Ö:<br>:Ö:Ï:Ö:<br>:Ö: ]$^-$ | [ O<br>\|<br>O–I–O<br>\|<br>O ]$^-$ |

59.

| | Ion | Valence Electrons | Electron Dot | Structural |
|---|---|---|---|---|
| (a) | $ClO^-$ | $7 + 6 + 1 = 14\ e^-$ | $[:\ddot{Cl}:\ddot{O}:]^-$ | $[Cl-O]^-$ |
| (b) | $BrO_2^-$ | $7 + 2(6) + 1 = 20\ e^-$ | $[:\ddot{O}:\ddot{Br}:\ddot{O}:]^-$ | $[O-Br-O]^-$ |
| (c) | $IO_3^-$ | $7 + 3(6) + 1 = 26\ e^-$ | $[:\ddot{O}:\ddot{I}:\ddot{O}:$ with $:\ddot{O}:$ below I$]^-$ | $[O-I-O$ with O bonded below I$]^-$ |
| (d) | $H_3O^+$ | $3(1) + 6 - 1 = 8\ e^-$ | $[H:\ddot{O}:H$ with H below O$]^+$ | $[H-O-H$ with H bonded below O$]^+$ |
| (e) | $ClO_4^-$ | $7 + 4(6) + 1 = 32\ e^-$ | $[:\ddot{O}:\ddot{Cl}:\ddot{O}:$ with $:\ddot{O}:$ above and below Cl$]^-$ | $[O-Cl-O$ with O bonded above and below Cl$]^-$ |

## Section 11.9 *Properties of Ionic and Molecular Compounds*

61.

| | Property | General Property of Ionic Compound |
|---|---|---|
| (a) | physical state | solid |
| (b) | structure | crystalline |
| (c) | melting point | high |
| (d) | boiling point | high |
| (e) | electrical conductivity | conductor |
| (f) | rate of reaction | fast |

63.

| | Property | Ionic or Molecular Compound |
|---|---|---|
| (a) | emerald crystal | ionic compound |
| (b) | greenish-yellow gas | molecular compound |
| (c) | melts at 801°C | ionic compound |
| (d) | boils at – 79°C | molecular compound |
| (e) | conductor in aq. solution | ionic compound |
| (f) | reacts slowly in solution | molecular compound |

## General Exercises

65. The radius of a sodium ion is less than the radius of a sodium atom because the ion has one less electron and the same number of protons as the atom. By having one less electron, the protons in the ion's nucleus have more attraction for the remaining electrons.

67. The properties of the elements Fe and $O_2$ are not related, either directly or indirectly, to the properties of the compound $Fe_2O_3$.

69.

| | Ions | | Ionic Formula |
|---|---|---|---|
| (a) | $Ca^{2+} + 2\ I^-$ | = | $CaI_2$ |
| (b) | $Ra^{2+} + O^{2-}$ | = | $RaO$ |
| (c) | $Ga^{3+} + 3\ F^-$ | = | $GaF_3$ |
| (d) | $3\ Ba^{2+} + 2\ P^{3-}$ | = | $Ba_3P_2$ |

71.

| | Ions | | Ionic Formula |
|---|---|---|---|
| (a) | $2\ Al^{3+} + 3\ CO_3^{2-}$ | = | $Al_2(CO_3)_3$ |
| (b) | $Sr^{2+} + 2\ OH^-$ | = | $Sr(OH)_2$ |
| (c) | $3\ Ag^+ + PO_4^{3-}$ | = | $Ag_3PO_4$ |
| (d) | $Cd^{2+} + 2\ NO_3^-$ | = | $Cd(NO_3)_2$ |

73. The properties of the elements $N_2$ and $O_2$ are not related to the properties of the compound $NO_2$.

75.

| | Molecule | Valence Electrons | Electron Dot | Structural |
|---|---|---|---|---|
| (a) | $SbH_3$ | $5 + 3(1) = 8\ e^-$ | H:S̈b:H<br>H (below Sb, shared pair) | H—Sb—H<br>\|<br>H |
| (b) | $GeCl_4$ | $4 + 4(7) = 32\ e^-$ | :C̈l:<br>:C̈l:G̈e:C̈l:<br>:C̈l: | Cl<br>\|<br>Cl—Ge—Cl<br>\|<br>Cl |

77. After the formation of chemical bonds, the valence electrons from individual atoms are delocalized throughout the entire molecule.

79.

| Molecule | Electron Dot Formulas |
|---|---|
| $SO_2$ | O::S:O: and :O:S::O (with lone pairs shown as dots) |

81.

| Molecule | Valence Electrons | Electron Dot | Structural |
|---|---|---|---|
| (1) $BF_3$ | $3 + 3(7) = 24\ e^-$ | :Cl: B :Cl: / :Cl: | Cl—B—Cl, Cl |
| (2) $PBr_5$ | $5 + 5(7) = 40\ e^-$ | :Br: :Br: / :Br :P: Br: / :Br: | Br, Br, Br—P—Br, Br |
| (3) $SF_6$ | $6 + 6(7) = 48\ e-$ | :F: / :F F: / :S: / :F F: / :F: | F, F, F, S, F, F, F |

83.

| Cation | Anion |
|---|---|
| $H^+$ | $H^-$ |

# The Gaseous State

CHAPTER 12

**Section 12.1** *Properties of Gases*

1. (1) Gases have an indefinite shape.
   (2) Gases may be expanded.
   (3) Gases may be compressed.
   (4) Gases have low densities.
   (5) Gases mix spontaneously to form homogeneous mixtures.

**Section 12.2** *Atmospheric Pressure and the Barometer*

3.

| | Units | Standard Atmospheric Pressure |
|---|---|---|
| (a) | atmospheres | 1.00 atm |
| (b) | millimeters of mercury | 760 mm Hg |
| (c) | torr | 760 torr |
| (d) | centimeters of mercury | 76.0 cm Hg |

5. (a) $5.25\ \cancel{\text{atm}} \times \dfrac{760\ \text{mm Hg}}{1.00\ \cancel{\text{atm}}} = 3990\ \text{mm Hg}$

(b) $5.25\ \cancel{\text{atm}} \times \dfrac{760\ \text{torr}}{1.00\ \cancel{\text{atm}}} = 3990\ \text{torr}$

(c) $5.25\ \cancel{\text{atm}} \times \dfrac{76.0\ \text{cm Hg}}{1.00\ \cancel{\text{atm}}} = 399\ \text{cm Hg}$

(d) $5.25\ \cancel{\text{atm}} \times \dfrac{29.9\ \text{in. Hg}}{1.00\ \cancel{\text{atm}}} = 157\ \text{in. Hg}$

7.* (a) $28.8\ \cancel{\text{in. Hg}} \times \dfrac{1.00\ \text{atm}}{29.9\ \cancel{\text{in. Hg}}} = 0.963\ \text{atm}$

(b) $28.8\ \cancel{\text{in. Hg}} \times \dfrac{760\ \text{mm Hg}}{29.9\ \cancel{\text{in. Hg}}} = 732\ \text{mm Hg}$

* continued on next page

7. (c) $28.8 \text{ \sout{in. Hg}} \times \dfrac{76.0 \text{ cm Hg}}{29.9 \text{ \sout{in. Hg}}} = 73.2 \text{ cm Hg}$

(d) $28.8 \text{ \sout{in. Hg}} \times \dfrac{760 \text{ torr}}{29.9 \text{ \sout{in. Hg}}} = 732 \text{ torr}$

**Section 12.3** *Vapor Pressure*

9. The vapor pressure of a liquid can be determined using a mercury barometer. A drop of liquid is introduced at the bottom of the barometer. The less dense liquid droplet floats to the top of the mercury and vaporizes. The vaporized liquid exerts a gas pressure, thus driving the mercury level down. The decrease in the mercury level corresponds to the vapor pressure of the liquid.

11. The general relationship between the vapor pressure of a liquid and its temperature is as follows: *As the temperature of a liquid increases, so does its vapor pressure.*

13. The temperature at which the vapor pressure is equal to the atmospheric pressure is the boiling point. Thus, the boiling point of acetone is 56°C because the vapor pressure at this temperature is 760 torr.

**Section 12.4** *Dalton's Law of Partial Pressures*

15. A gas is collected over water to determine its volume. A volume of gas displaces an equal volume of water (refer to Figures 2.3 and 12.7). When collecting a gas over water, it is usually assumed that the gas is not appreciably soluble in water. This may not be true as some gases, such as carbon dioxide, are very soluble in water.

17. $P_{\text{water vapor}} = 24 \text{ torr}$
$P_{\text{total}} = 766 \text{ torr}$
$P_{\text{water vapor}} + P_{\text{oxygen}} = P_{\text{total}}$
$P_{\text{oxygen}} = P_{\text{total}} - P_{\text{water vapor}} = 766 \text{ torr} - 24 \text{ torr} = 742 \text{ torr}$

19. $P_{\text{nitrogen}} = 587 \text{ torr}$
$P_{\text{oxygen}} = 158 \text{ torr}$
$P_{\text{argon}} = 7 \text{ torr}$
$P_{\text{atmosphere}} = P_{\text{nitrogen}} + P_{\text{oxygen}} + P_{\text{argon}} = 752 \text{ torr}$

**Section 12.5** *Variables Affecting Gas Pressure*

21. (1) Volume
(2) Temperature
(3) Number of molecules

23.

| | Change | Result | Explanation |
|---|---|---|---|
| (a) | volume increases | pressure decreases | molecules are farther apart and collide less frequently |
| (b) | temperature increases | pressure increases | molecules move faster and collide with a greater frequency and energy |
| (c) | moles of gas increase | pressure increases | more molecules cause more collisions |

25.

| | Change | Result |
|---|---|---|
| (a) | volume increases | pressure decreases |
| (b) | temperature increases | pressure increases |
| (c) | moles of gas decrease | pressure decreases |

**Section 12.6** *Boyle's Law: Pressure / Volume Changes*

27. Pressure vs. Volume

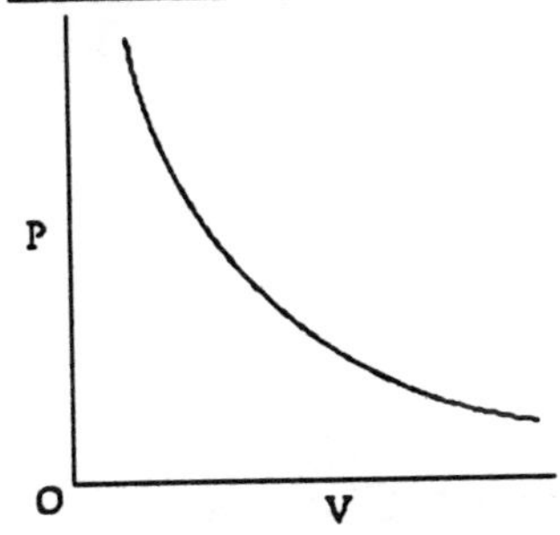


29. $P_1 \times V_{factor} = P_2$

$$0.750 \text{ atm} \times \frac{250.0 \text{ mL}}{655.0 \text{ mL}} = 0.286 \text{ atm}$$

31. $P_1 \times V_{factor} = P_2$

$$15.0 \text{ psi} \times \frac{50.0 \text{ mL}}{44.0 \text{ mL}} = 17.0 \text{ psi}$$

**Section 12.7** *Charles' Law: Volume / Temperature Changes*

33. Volume vs. Temperature

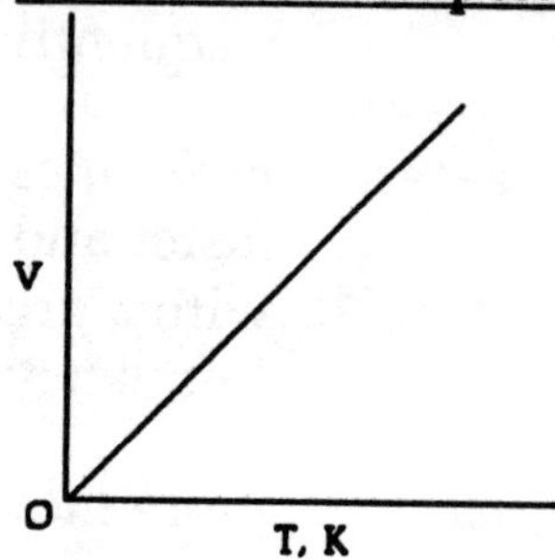


35. $V_1 \times T_{factor} = V_2$

$25°C + 273 = 298\ K$
$50°C + 273 = 323\ K$

$$335\ mL \times \frac{323\ \cancel{K}}{298\ \cancel{K}} = 363\ mL\ O_2$$

37. $V_1 \times T_{factor} = V_2$

$0°C + 273 = 273\ K$
$100°C + 273 = 373\ K$

$$80.0\ cm^3 \times \frac{373\ \cancel{K}}{273\ \cancel{K}} = 109\ cm^3\ F_2$$

**Section 12.8** *Gay-Lussac's Law: Pressure / Temperature Changes*

39. Pressure vs. Temperature

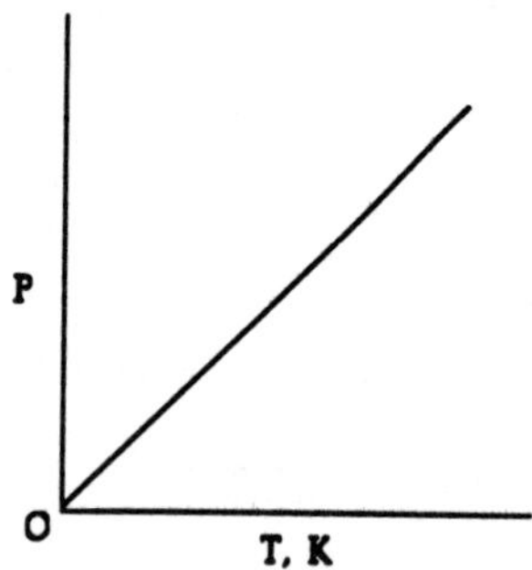

41. $P_1 \times T_{factor} = P_2$

$20°C + 273 = 293\ K$
$200°C + 273 = 473\ K$

$$760\ torr \times \frac{473\ \cancel{K}}{293\ \cancel{K}} = 1230\ torr$$

43. $P_1 \times T_{factor} = P_2$

$0°C + 273 = 273\ K$
$100°C + 273 = 373\ K$

$$76.0\ cm\ Hg \times \frac{373\ \cancel{K}}{273\ \cancel{K}} = 104\ cm\ Hg$$

**Section 12.9** *Combined Gas Law*

45.

| | P | V | T |
|---|---|---|---|
| initial | 772 mm Hg | 100.0 mL | 21°C + 273 = 294 K |
| final | 760 mm Hg | $V_2$ | 273 K |

$V_1 \times P_{factor} \times T_{factor} = V_2$

$$100.0\ mL \times \frac{772\ \cancel{mm\ Hg}}{760\ \cancel{mm\ Hg}} \times \frac{273\ \cancel{K}}{294\ \cancel{K}} = 94.3\ mL\ H_2$$

47.

| | P | V | T |
|---|---|---|---|
| initial | 760 torr | 2.00 L | 273 K |
| final | 365 torr | $V_2$ | 75°C + 273 = 348 K |

$V_1 \times P_{factor} \times T_{factor} = V_2$

$$2.00\ L \times \frac{760\ \cancel{torr}}{365\ \cancel{torr}} \times \frac{348\ \cancel{K}}{273\ \cancel{K}} = 5.31\ L\ air$$

49.

| | P | V | T |
|---|---|---|---|
| initial | 760 torr | 1250 mL | 273 K |
| final | $P_2$ | 255 mL | 300°C + 273 = 573 K |

$$P_1 \times V_{factor} \times T_{factor} = P_2$$

$$760 \text{ torr} \times \frac{1250 \text{ mL}}{255 \text{ mL}} \times \frac{573 \text{ K}}{273 \text{ K}} = 7820 \text{ torr}$$

51.

| | P | V | T |
|---|---|---|---|
| initial | 225 mm Hg | 500.0 mL | –125°C + 273 = 148 K |
| final | $P_2$ | 220.0 mL | 100°C + 273 = 373 K |

$$P_1 \times V_{factor} \times T_{factor} = P_2$$

$$225 \text{ mm Hg} \times \frac{500.0 \text{ mL}}{220.0 \text{ mL}} \times \frac{373 \text{ K}}{148 \text{ K}} = 1290 \text{ mm Hg}$$

53.

| | P | V | T |
|---|---|---|---|
| initial | 760 torr | 50.0 mL | 273 K |
| final | 350 torr | 350.0 mL | $T_2$ |

$$T_1 \times P_{factor} \times V_{factor} = T_2$$

$$273 \text{ K} \times \frac{350 \text{ torr}}{760 \text{ torr}} \times \frac{350.0 \text{ mL}}{50.0 \text{ mL}} = 880 \text{ K}$$

$$880 \text{ K} - 273 = 607°\text{C}$$

## Section 12.10 *Ideal Gas Behavior*

55. Characteristics of an Ideal Gas

(1) Gases are made up of extremely tiny molecules.
(2) Gas molecules demonstrate rapid motion, move in straight lines, and travel in random directions.
(3) Gas molecules show no attraction for one another.
(4) Gas molecules have elastic collisions.
(5) The average kinetic energy of gas molecules is proportional to the Kelvin temperature.

57. A real gas behaves most like an ideal gas at high temperatures and low pressures.

59.

| | Description | Gas | | Description | Gas |
|---|---|---|---|---|---|
| (a) | most kinetic energy | all equal | (b) | least kinetic energy | all equal |
| (c) | highest velocity | He | (d) | lowest velocity | Ar |

(Note: Since the gases are at the same temperature they have the same kinetic energy.)

61. An ideal gas at 0 K exerts no pressure (0 mm Hg).

**Section 12.11** *Ideal Gas Equation*

63. $P = \frac{n\,R\,T}{V}$

$n = 0.500 \text{ mol } H_2$

$T = 25°C + 273 = 298 \text{ K}$

$V = 50.0 \cancel{\text{mL}} \times \frac{1 \text{ L}}{1000 \cancel{\text{mL}}} = 0.0500 \text{ L}$

$$P = \frac{0.500 \cancel{\text{mol}} \times 298 \cancel{\text{K}}}{0.0500 \cancel{\text{L}}} \times \frac{0.0821 \text{ atm} \cdot \cancel{\text{L}}}{1 \cancel{\text{mol}} \cdot \cancel{\text{K}}} = 245 \text{ atm}$$

65. $n = \frac{P\,V}{R\,T}$

$P = 125 \cancel{\text{psi}} \times \frac{1.00 \text{ atm}}{14.7 \cancel{\text{psi}}} = 8.50 \text{ atm}$

$V = 10.0 \text{ L}$

$T = 373 \text{ K}$

$$n = \frac{8.50 \cancel{\text{atm}} \times 10.0 \cancel{\text{L}}}{373 \cancel{\text{K}}} \times \frac{1 \text{ mol} \cdot \cancel{\text{K}}}{0.0821 \cancel{\text{atm}} \cdot \cancel{\text{L}}} = 2.78 \text{ mol } N_2O$$

67. $MM = \frac{g\,R\,T}{P\,V}$

$g = 2.14 \text{ g}$

$V = 1 \text{ L}$

$P = 1.00 \text{ atm}$

$T = 273 \text{ K}$

$$MM = \frac{2.14 \text{ g} \times 273 \cancel{\text{K}}}{1.00 \cancel{\text{atm}} \times 1.00 \cancel{\text{L}}} \times \frac{0.0821 \cancel{\text{atm}} \cdot \cancel{\text{L}}}{1 \text{ mol} \cdot \cancel{\text{K}}} = 48.0 \text{ g/mol}$$

69. $n = \frac{PV}{RT}$

$P = 710 \text{ torr} \times \frac{1.00 \text{ atm}}{760 \text{ torr}} = 0.934 \text{ atm}$

$V = 1550 \text{ mL} \times \frac{1 \text{ L}}{1000 \text{ mL}} = 1.55 \text{ L}$

$T = 20°\text{C} + 273 = 293 \text{ K}$

$n = \frac{0.934 \cancel{\text{atm}} \times 1.55 \cancel{\text{L}}}{293 \cancel{\text{K}}} \times \frac{1 \text{ mol} \cdot \cancel{\text{K}}}{0.0821 \cancel{\text{atm}} \cdot \cancel{\text{L}}} = 0.0602 \text{ mol } Cl_2$

$0.0602 \cancel{\text{mol } Cl_2} \times \frac{71.0 \text{ g } Cl_2}{1 \cancel{\text{mol } Cl_2}} = 4.27 \text{ g } Cl_2$

71. $\frac{0.0821 \cancel{\text{atm}} \cdot \text{L}}{1 \text{ mol} \cdot \text{K}} \times \frac{760 \text{ torr}}{1.00 \cancel{\text{atm}}} = \frac{62.4 \text{ torr} \cdot \text{L}}{\text{mol} \cdot \text{K}}$

## General Exercises

73. $2500 \cancel{\text{in.}^2} \times \frac{14.7 \text{ lbs}}{\cancel{\text{in.}^2}} = 37{,}000 \text{ lbs}$

75. A liquid boils when its vapor pressure equals the external pressure. When the vacuum pump lowers the external pressure to that of the vapor pressure of the water, the water begins to boil.

77. From Figure 12.6: The boiling point of ethyl alcohol at 0.5 atm is ~ 60°C.

79. $P_{total} = 764 \text{ mm Hg}$

$P_{water\ vapor} = 19.8 \text{ mm Hg}$

$P_{oxygen} + P_{water\ vapor} = P_{total}$

$P_{oxygen} = P_{total} - P_{water\ vapor}$

$P_{oxygen} = 764 \text{ mm Hg} - 19.8 \text{ mm Hg} = 744.2 \text{ mm Hg}$

| | P | V | T |
|---|---|---|---|
| initial | 744.2 mm Hg | 42.5 mL | 22°C + 273 = 295 K |
| final | 760 mm Hg | $V_2$ | 273 K |

$V_1 \times P_{factor} \times T_{factor} = V_2$

$42.5 \text{ mL} \times \frac{744.2 \cancel{\text{mm Hg}}}{760 \cancel{\text{mm Hg}}} \times \frac{273 \cancel{\text{K}}}{295 \cancel{\text{K}}} = 38.5 \text{ mL } O_2$

81. $P = \dfrac{nRT}{V}$

$$n = 1.51 \times 10^{24}\ \cancel{\text{atoms Kr}} \times \frac{1\ \text{mol Kr}}{6.02 \times 10^{23}\ \cancel{\text{atoms Kr}}} = 2.51\ \text{mol Kr}$$

$T = 25°C + 273 = 298\ K$
$V = 5.00\ L$

$$P = \frac{2.51\ \cancel{\text{mol}} \times 298\ \cancel{\text{K}}}{5.00\ \cancel{\text{L}}} \times \frac{0.0821\ \text{atm} \cdot \cancel{\text{L}}}{1\ \cancel{\text{mol}} \cdot \cancel{\text{K}}} = 12.3\ \text{atm}$$

83. $V = \dfrac{nRT}{P}$

$$n = 3.38 \times 10^{22}\ \cancel{\text{molecules NO}} \times \frac{1\ \text{mol NO}}{6.02 \times 10^{23}\ \cancel{\text{molecules NO}}}$$
$$= 0.0561\ \text{mol NO}$$

$T = 100°C + 273 = 373\ K$

$$P = 255\ \cancel{\text{torr}} \times \frac{1.00\ \text{atm}}{760\ \cancel{\text{torr}}} = 0.336\ \text{atm}$$

$$V = \frac{0.0561\ \cancel{\text{mol}} \times 373\ \cancel{\text{K}}}{0.336\ \cancel{\text{atm}}} \times \frac{0.0821\ \cancel{\text{atm}} \cdot \text{L}}{1\ \cancel{\text{mol}} \cdot \cancel{\text{K}}} = 5.11\ \text{L NO}$$

85. The number of molecules in a closed system remains constant because there is no way to add more molecules without opening the system.

87. Scuba divers breathe compressed air that is ~80% nitrogen and ~ 20% oxygen. Deep sea divers breathe a mixture that is mostly helium. Since He is lighter than $N_2$, the helium atoms move faster than nitrogen molecules and produce higher pitched voices.

# Liquids, Solids, and Water

## Section 13.1 *The Liquid State*

1. General Properties of Liquids
   (1) Liquids have an indefinite shape but a fixed volume.
   (2) Liquids usually flow readily.
   (3) Liquids do not expand or compress to any degree.
   (4) Liquids have a high density compared to gases.
   (5) Liquids that are soluble mix uniformly.

3.

| | Substance | Temperature | Physical State |
|---|---|---|---|
| (a) | $H_2O$ | – 20.0°C | solid |
| (b) | $H_2O$ | 120.0°C | gas |
| (c) | $NH_3$ | – 195.0°C | solid |
| (d) | $NH_3$ | 0.0°C | gas |
| (e) | $CHCl_3$ | – 55.5°C | liquid |
| (f) | $CHCl_3$ | 100.0°C | gas |

## Section 13.2 *Properties of Liquids*

5. Water molecules in the liquid state evaporate to the gaseous state and form water vapor. Simultaneously, some water molecules in the vapor condense back to the liquid. By definition, the vapor pressure of water is the total pressure exerted by the water molecules in the vapor when the rate of evaporation is equal to the rate of condensation.

7. The viscosity of a liquid is a measure of its resistance to flow. Since the intermolecular attraction is high in water, the viscosity of water is higher than other molecules of the same size.

9.

| | Property of Water | Value | | Property of Water | Value |
|---|---|---|---|---|---|
| (a) | vapor pressure | low | (b) | boiling point | high |
| (c) | viscosity | high | (d) | surface tension | high |

## Section 13.3 *Intermolecular Forces*

11.

| | Molecule | Strongest Intermolecular Force |
|---|---|---|
| (a) | $CH_3–CH_2–CH_2–CH_2–CH_2–CH_3$ | dispersion forces |
| (b) | $CH_3–OH$ | hydrogen bonding |
| (c) | $CH_3–Cl$ | dipole forces |
| (d) | $CH_3–O–CH_3$ | dipole forces |

13.

| | Molecules | Higher Vapor Pressure |
|---|---|---|
| (a) | $CH_3COOH$ or $C_2H_5Cl$ | $C_2H_5Cl$ |
| (b) | $C_2H_5OH$ or $CH_3OCH_3$ | $CH_3OCH_3$ |

15.

| | Molecules | Higher Viscosity |
|---|---|---|
| (a) | $CH_3COOH$ or $C_2H_5Cl$ | $CH_3COOH$ |
| (b) | $C_2H_5OH$ or $CH_3OCH_3$ | $C_2H_5OH$ |

## Section 13.4 *The Solid State*

17. General Properties of Solids
    (1) Solids have a definite shape and a fixed volume.
    (2) Solids are either crystalline or noncrystalline.
    (3) Solids do not compress or expand to any degree.
    (4) Solids usually have a slightly higher density than their corresponding liquids.
    (5) Solids do not mix by diffusion.

19.

| | Substance | Temperature | Physical State |
|---|---|---|---|
| (a) | Ga | 0°C | solid |
| (b) | Ga | 100°C | liquid |
| (c) | Sn | 0°C | solid |
| (d) | Sn | 100°C | solid |
| (e) | Hg | 0°C | liquid |
| (f) | Hg | 100°C | liquid |

## Section 13.5 *Crystalline Solids*

21. (1) Ionic solid: $CuSO_4$
    (2) Metallic solid: Ag
    (3) Molecular solid: $I_2$

23.

| | Crystalline Solid | Type of Particles |
|---|---|---|
| (a) | ionic solid | ions |
| (b) | molecular solid | molecules |
| (c) | metallic solid | metal atoms |

25. | | Crystalline Solid | Classification
---|---|---|---
| (a) | Zn | metallic solid
| (b) | ZnO | ionic solid
| (c) | $P_4$ | molecular solid
| (d) | IBr | molecular solid

**Section 13.6** *Changes of Physical State*

27. Heating of Ethanol

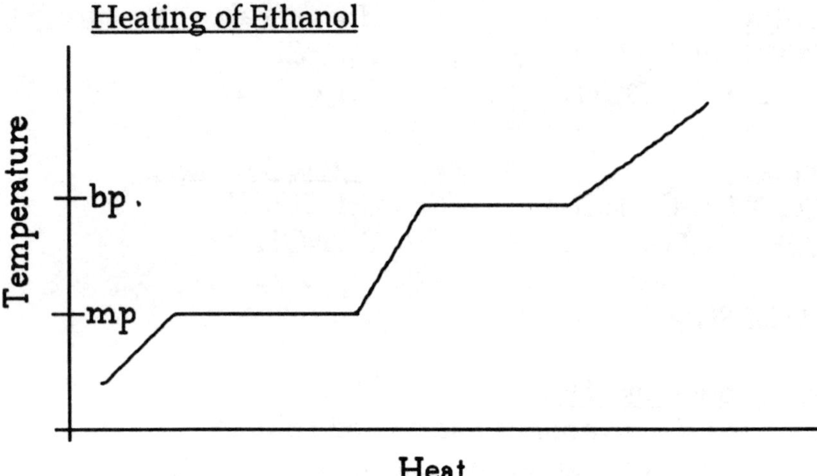


29. Energy required to melt ice:

$$125\ \cancel{g} \times \frac{80.0\ \text{cal}}{1\ \cancel{g}} = 1.00 \times 10^4\ \text{cal}\ \ (10.0\ \text{kcal})$$

31. Energy required to heat water:

$$25.0\ \cancel{g} \times \frac{1.00\ \text{cal}}{1\ \cancel{g} \cdot \cancel{°C}} \times (100.0 - 25.0)\cancel{°C} = 1{,}880\ \text{cal}\ \ (1.88\ \text{kcal})$$

Energy required to vaporize water:

$$25.0\ \cancel{g} \times \frac{540\ \text{cal}}{1\ \cancel{g}} = 13{,}500\ \text{cal}\ \ (13.5\ \text{kcal})$$

Total heat energy required:
1,880 cal + 13,500 cal = 15,400 cal (15.4 kcal)

33.* Energy required to melt ice:

$$115\ \cancel{g} \times \frac{80.0\ \text{cal}}{1\ \cancel{g}} = 9{,}200\ \text{cal}\ \ (9.20\ \text{kcal})$$

Energy required to heat water:

$$115\ \cancel{g} \times \frac{1.00\ \text{cal}}{1\ \cancel{g} \cdot \cancel{°C}} \times (100.0 - 0.0)\cancel{°C} = 11{,}500\ \text{cal}\ \ (11.5\ \text{kcal})$$

---

* continued on next page

33. Energy required to vaporize water:

$$115\ \cancel{g} \times \frac{540\ \text{cal}}{1\ \cancel{g}} = 62{,}100\ \text{cal}\ \ (62.1\ \text{kcal})$$

Total heat energy required:
9,200 cal + 11,500 cal + 62,100 cal = 82,800 cal (82.8 kcal)

35. Energy required to heat ice:

$$38.5\ \cancel{g} \times \frac{0.50\ \text{cal}}{1\ \cancel{g} \cdot \cancel{^{\circ}C}} \times [0.0 - (-20.0)]\cancel{^{\circ}C} = 385\ \text{cal}\ \ (0.385\ \text{kcal})$$

Energy required to melt ice:

$$38.5\ \cancel{g} \times \frac{80.0\ \text{cal}}{1\ \cancel{g}} = 3{,}080\ \text{cal}\ \ (3.08\ \text{kcal})$$

Energy required to heat water:

$$38.5\ \cancel{g} \times \frac{1.00\ \text{cal}}{1\ \cancel{g} \cdot \cancel{^{\circ}C}} \times (100.0 - 0.0)\cancel{^{\circ}C} = 3850\ \text{cal}\ \ (3.85\ \text{kcal})$$

Energy required to vaporize water:

$$38.5\ \cancel{g} \times \frac{540\ \text{cal}}{1\ \cancel{g}} = 20{,}800\ \text{cal}\ \ (20.8\ \text{kcal})$$

Total heat energy required:
385 cal + 3,080 cal + 3850 cal + 20,800 cal = 28,100 cal (28.1 kcal)

37.* Energy required to heat ice:

$$100.0\ \cancel{g} \times \frac{0.50\ \text{cal}}{1\ \cancel{g} \cdot \cancel{^{\circ}C}} \times [0.0 - (-40.0)]\cancel{^{\circ}C} = 2{,}000\ \text{cal}\ \ (2.00\ \text{kcal})$$

Energy required to melt ice:

$$100.0\ \cancel{g} \times \frac{80.0\ \text{cal}}{1\ \cancel{g}} = 8{,}000\ \text{cal}\ \ (8.00\ \text{kcal})$$

Energy required to heat water:

$$100.0\ \cancel{g} \times \frac{1.00\ \text{cal}}{1\ \cancel{g} \cdot \cancel{^{\circ}C}} \times (100.0 - 0.0)\cancel{^{\circ}C} = 10{,}000\ \text{cal}\ \ (10.0\ \text{kcal})$$

Energy required to vaporize water:

$$100.0\ \cancel{g} \times \frac{540\ \text{cal}}{1\ \cancel{g}} = 54{,}000\ \text{cal}\ \ (54.0\ \text{kcal})$$

---

* continued on next page

37. Energy required to heat steam:

$$100.0\ \cancel{g} \times \frac{0.48\ \text{cal}}{1\ \cancel{g} \cdot \cancel{^\circ C}} \times (125.0 - 100.0)\cancel{^\circ C} = 1{,}200\ \text{cal}\ \ (1.20\ \text{kcal})$$

Total heat energy required:

$$2{,}000\ \text{cal} + 8{,}000\ \text{cal} + 10{,}000\ \text{cal} + 54{,}000\ \text{cal} + 1{,}200\ \text{cal} = 75{,}200\ \text{cal} = (75.2\ \text{kcal})$$

**Section 13.7** *The Water Molecule*

39.

| Molecule | Electron Dot Formula | Structural Formula |
|---|---|---|
| $H_2O$ | $H:\ddot{\underset{\cdot\cdot}{O}}:H$ | H—O—H |

41. The observed bond angle in a water molecule: 104.5°

43. Structural Formula With Delta Notation

```
      δ−
      O
     / \
    H   H
   δ+   δ+
```

45. Hydrogen Bonding in Hydrogen Fluoride

```
H—F·········H—F
        ↑
      H-bond
```

**Section 13.8** *Physical Properties of Water*

47. An "ammonia ice cube" *floats* because solid ammonia is less dense than liquid ammonia.

49.

| | Molecules | Higher Melting Point |
|---|---|---|
| (a) | $H_2O$ or $H_2S$ | $H_2O$ |
| (b) | $H_2S$ or $H_2Se$ | $H_2Se$ |

51.

| | Molecules | Higher Heat of Fusion |
|---|---|---|
| (a) | $H_2O$ or $H_2S$ | $H_2O$ |
| (b) | $H_2S$ or $H_2Se$ | $H_2Se$ |

53.

| | Property | As Molar Mass Increases |
|---|---|---|
| (a) | melting point | increases |
| (b) | boiling point | increases |
| (c) | heat of fusion | increases |
| (d) | heat of vaporization | increases |

55. 

| Similarities | Differences |
|---|---|
| Both are colorless | Heavy water is toxic to animals and humans |
| Both are odorless | Molar mass |
| Both are tasteless | Density |
| | Boiling and melting points |
| | Heat of fusion |

## Section 13.9 *Chemical Properties of Water*

57. $2\ H_2O_{(l)} \xrightarrow{\text{electricity}} 2\ H_{2(g)} + O_{2(g)}$

59. (a) $2\ Li_{(s)} + 2\ H_2O_{(l)} \rightarrow 2\ LiOH_{(aq)} + H_{2(g)}$
(b) $2\ Rb_{(s)} + 2\ H_2O_{(l)} \rightarrow 2\ RbOH_{(aq)} + H_{2(g)}$
(c) $Na_2O_{(s)} + H_2O_{(l)} \rightarrow 2\ NaOH_{(aq)}$
(d) $Cs_2O_{(s)} + H_2O_{(l)} \rightarrow 2\ CsOH_{(aq)}$
(e) $CO_{2(g)} + H_2O_{(l)} \rightarrow H_2CO_{3(aq)}$
(f) $P_2O_{5(s)} + 3\ H_2O_{(l)} \rightarrow 2\ H_3PO_{4(aq)}$

61. (a) $2\ C_3H_{6(g)} + 9\ O_{2(g)} \xrightarrow{\text{spark}} 6\ CO_{2(g)} + 6\ H_2O_{(g)}$
(b) $C_3H_6O_{(g)} + 4\ O_{2(g)} \xrightarrow{\text{spark}} 3\ CO_{2(g)} + 3\ H_2O_{(g)}$
(c) $2\ HF_{(aq)} + Ca(OH)_{2(aq)} \rightarrow CaF_{2(aq)} + 2\ HOH_{(l)}$
(d) $H_2CO_{3(aq)} + 2\ KOH_{(aq)} \rightarrow K_2CO_{3(aq)} + 2\ HOH_{(l)}$
(e) $Na_2Cr_2O_7 \cdot 2\ H_2O_{(s)} \xrightarrow{\Delta} Na_2Cr_2O_{7(s)} + 2\ H_2O_{(g)}$
(f) $Ca(NO_3)_2 \cdot 4\ H_2O_{(s)} \xrightarrow{\Delta} Ca(NO_3)_{2(s)} + 4\ H_2O_{(g)}$

## Section 13.10 *Hydrates*

63.

| | Compound | Systematic Name |
|---|---|---|
| (a) | $MgSO_4 \cdot 7\ H_2O$ | magnesium sulfate heptahydrate |
| (b) | $Co(CN)_3 \cdot 3\ H_2O$ | cobalt(III) cyanide trihydrate |
| (c) | $MnSO_4 \cdot H_2O$ | manganese(II) sulfate monohydrate |
| (d) | $Na_2Cr_2O_7 \cdot 2\ H_2O$ | sodium dichromate dihydrate |
| (e) | $Sr(NO_3)_2 \cdot 6\ H_2O$ | strontium nitrate hexahydrate |
| (f) | $Co(C_2H_3O_2)_2 \cdot 4\ H_2O$ | cobalt(II) acetate tetrahydrate |
| (g) | $CuSO_4 \cdot 5\ H_2O$ | copper(II) sulfate pentahydrate |
| (h) | $Cr(NO_3)_3 \cdot 9\ H_2O$ | chromium(III) nitrate nonahydrate |

65. (a) Percentage water in $SrCl_2 \cdot 6\ H_2O$

MM of $SrCl_2$ = 87.6 g + 2(35.5) g = 158.6 g

$$\text{Percentage water: } \frac{6(18.0)\text{ g}}{158.6\text{ g} + 6(18.0)\text{ g}} \times 100 = 40.5\%\ H_2O$$

(b) Percentage water in $K_2Cr_2O_7 \cdot 2\ H_2O$

MM of $K_2Cr_2O_7$ = 2(39.1) g + 2(52.0) g + 7(16.0) g = 294.2 g

$$\text{Percentage water: } \frac{2(18.0)\text{ g}}{294.2\text{ g} + 2(18.0)\text{ g}} \times 100 = 10.9\%\ H_2O$$

(c) Percentage water in $MgSO_4 \cdot 7\ H_2O$

MM of $MgSO_4$ = 24.3 g + 32.1 g + 4(16.0) g = 120.4 g

$$\text{Percentage water: } \frac{7(18.0)\text{ g}}{120.4\text{ g} + 7(18.0)\text{ g}} \times 100 = 51.1\%\ H_2O$$

(d) Percentage water in $Co(CN)_3 \cdot 3\ H_2O$

MM of $Co(CN)_3$ = 58.9 g + 3(12.0) g + 3(14.0) g = 136.9 g

$$\text{Percentage water: } \frac{3(18.0)\text{ g}}{136.9\text{ g} + 3(18.0)\text{ g}} \times 100 = 28.3\%\ H_2O$$

(e) Percentage water in $MnSO_4 \cdot H_2O$

MM of $MnSO_4$ = 54.9 g + 32.1 g + 4(16.0) g = 151.0 g

$$\text{Percentage water: } \frac{18.0\text{ g}}{151.0\text{ g} + 18.0\text{ g}} \times 100 = 10.7\%\ H_2O$$

(f) Percentage water in $Na_2CrO_4 \cdot 4\ H_2O$

MM of $Na_2CrO_4$ = 2(23.0) g + 52.0 g + 4(16.0) g = 162.0 g

$$\text{Percentage water: } \frac{4(18.0)\text{ g}}{162.0\text{ g} + 4(18.0)\text{ g}} \times 100 = 30.8\%\ H_2O$$

67.* (a) $NiCl_2 \cdot X\,H_2O_{(s)} \xrightarrow{\Delta} NiCl_{2(s)} + X\,H_2O_{(g)}$

$$21.7\ \cancel{g\ H_2O} \times \frac{1\ mol\ H_2O}{18.0\ \cancel{g\ H_2O}} = 1.21\ mol\ H_2O$$

$$78.3\ \cancel{g\ NiCl_2} \times \frac{1\ mol\ NiCl_2}{129.7\ \cancel{g\ NiCl_2}} = 0.604\ mol\ NiCl_2$$

$$NiCl_2 \cdot \frac{1.21}{0.604} H_2O \qquad \frac{1.21}{0.604} = 2.00$$

Formula: $NiCl_2 \cdot 2\,H_2O$

(b) $Sr(NO_3)_2 \cdot X\,H_2O_{(s)} \xrightarrow{\Delta} Sr(NO_3)_{2(s)} + X\,H_2O_{(g)}$

$$33.8\ \cancel{g\ H_2O} \times \frac{1\ mol\ H_2O}{18.0\ \cancel{g\ H_2O}} = 1.88\ mol\ H_2O$$

$$66.2\ \cancel{g\ Sr(NO_3)_2} \times \frac{1\ mol\ Sr(NO_3)_2}{211.6\ \cancel{g\ Sr(NO_3)_2}} = 0.313\ mol\ Sr(NO_3)_2$$

$$Sr(NO_3)_2 \cdot \frac{1.88}{0.313} H_2O \qquad \frac{1.88}{0.313} = 6.01$$

Formula: $Sr(NO_3)_2 \cdot 6\,H_2O$

(c) $CrI_3 \cdot X\,H_2O_{(s)} \xrightarrow{\Delta} CrI_{3(s)} + X\,H_2O_{(g)}$

$$27.2\ \cancel{g\ H_2O} \times \frac{1\ mol\ H_2O}{18.0\ \cancel{g\ H_2O}} = 1.51\ mol\ H_2O$$

$$72.8\ \cancel{g\ CrI_3} \times \frac{1\ mol\ CrI_3}{432.7\ \cancel{g\ CrI_3}} = 0.168\ mol\ CrI_3$$

$$CrI_3 \cdot \frac{1.51}{0.168} H_2O \qquad \frac{1.51}{0.168} = 8.99$$

Formula: $CrI_3 \cdot 9\,H_2O$

* continued on next page

67. (d) $Ca(NO_3)_2 \cdot X\,H_2O_{(s)} \xrightarrow{\Delta} Ca(NO_3)_{2(s)} + X\,H_2O_{(g)}$

$$30.5\ \cancel{g\ H_2O} \times \frac{1\ mol\ H_2O}{18.0\ \cancel{g\ H_2O}} = 1.69\ mol\ H_2O$$

$$69.5\ \cancel{g\ Ca(NO_3)_2} \times \frac{1\ mol\ Ca(NO_3)_2}{164.1\ \cancel{g\ Ca(NO_3)_2}} = 0.424\ mol\ Ca(NO_3)_2$$

$$Ca(NO_3)_2 \cdot \frac{1.69}{0.424} H_2O \qquad \frac{1.69}{0.424} = 3.99$$

Formula: $Ca(NO_3)_2 \cdot 4\,H_2O$

## Section 13.11 *Water Purification*

69.

| | Type of Water | Cations |
|---|---|---|
| (a) | hard water | $Ca^{2+}$, $Mg^{2+}$, $Fe^{3+}$, $Na^{+}$ |
| (b) | soft water | $Na^{+}$ |
| (c) | deionized water | negligible |
| (d) | distilled water | negligible |

71. Cations found in tap water ($Ca^{2+}$, $Mg^{2+}$, $Fe^{3+}$) react with soap anions to form a precipitate.

## General Exercises

73. Approximate Percent of Water on Earth
   (a) salt water: ~ 97%
   (b) nonfrozen fresh water: ~ 1%

75. A homogeneous liquid mixture is a liquid that contains one or more dissolved substances; for example, ethyl alcohol dissolved in water.

77. Sulfuric acid, $H_2SO_4$, is quite polar so we can predict that it behaves somewhat like water. That is, owing to surface tension sulfuric acid "rain drops" on Venus will form spheres as rain drops do on earth.

79. Energy released when ethylene glycol cools:

$$1250\ \cancel{g} \times \frac{0.561\ \text{cal}}{1\ \cancel{g} \cdot \cancel{°C}} \times [-(-11.5 - 25.0)]\cancel{°C} = 25{,}600\ \text{cal}\ \ (25.6\ \text{kcal})$$

Energy released when ethylene glycol solidifies:

$$1250\ \cancel{g} \times \frac{43.3\ \text{cal}}{1\ \cancel{g}} = 54{,}100\ \text{cal}\ \ (54.1\ \text{kcal})$$

Total heat energy released:
25,600 cal + 54,100 cal = 79,700 cal (79.7 kcal)

81. Physical Properties of Water Without H-Bonding

mp: $-85.5 + [-85.5 - (-60.4)] = \sim -110°C$
bp: $-60.7 + [-60.7 - (-41.5)] = \sim -80°C$
$H_{fusion}$: $568 - (599 - 568) = \sim 540$ cal/mol
$H_{vapor}$: $4450 - (4620 - 4450) = \sim 4300$ cal/mol

(Note: These values are estimated based on the interval between $H_2S$ and $H_2Se$.)

83. $Mg^{2+}{}_{(aq)} + 2\ Na(soap)_{(aq)} \rightarrow Mg(soap)_{2(s)} + 2\ Na^{+}{}_{(aq)}$

# CHAPTER 14

# Solutions

## Section 14.1 *Gases in a Liquid Solution*

1.

| | Change | Result |
|---|---|---|
| (a) | temperature of solution decreases | solubility of ammonia increases |
| (b) | partial pressure of $NH_3$ increases | solubility of ammonia increases |

3. standard solubility × pressure factor = new solubility

$$\frac{1.45 \text{ g } CO_2}{1 \text{ L champagne}} \times \frac{10.0 \cancel{\text{atm}}}{1.00 \cancel{\text{atm}}} = 14.5 \text{ g } CO_2/1 \text{ L champagne}$$

5. $$\frac{0.63 \text{ g } Cl_2}{100 \text{ g } H_2O} \times \frac{1200 \cancel{\text{mm Hg}}}{760 \cancel{\text{mm Hg}}} = 0.99 \text{ g } Cl_2/100 \text{ g } H_2O$$

## Section 14.2 *Liquids in a Liquid Solution*

7. (a) polar solvent + polar solvent = miscible
   (b) polar solvent + nonpolar solvent = immiscible

9.

| Polar Solvents | Nonpolar Solvents |
|---|---|
| (a) $H_2O$ | (b) $C_6H_{14}$ |
| (c) $C_3H_6O$ | (d) $CHCl_3$ |

11.

| | Solvent | Miscible or Immiscible with Water |
|---|---|---|
| (a) | $C_7H_{16}$ | immiscible |
| (b) | $CH_3OH$ | miscible |
| (c) | $C_4H_8O$ | miscible |
| (d) | $C_7H_8$ | immiscible |

13. Add several drops of the unknown liquid into a test tube containing water which is a polar solvent. If the unknown liquid is also polar, it will be miscible with the water. If the liquid is nonpolar, it will be immiscible and two distinct layers will form in the test tube.

## Section 14.3 *Solids in a Liquid Solution*

15. (a) polar solute + polar solvent = soluble
    (b) nonpolar solute + polar solvent = insoluble
    (c) ionic solute + polar solvent = soluble

17.

| | Compound | Formula | Solubility in Water |
|---|---|---|---|
| (a) | naphthalene | $C_{10}H_8$ | insoluble |
| (b) | potassium hydroxide | $KOH$ | soluble |
| (c) | calcium acetate | $Ca(C_2H_3O_2)_2$ | soluble |
| (d) | trichlorotoluene | $C_7H_5Cl_3$ | insoluble |
| (e) | glycine | $C_2H_5NO_2$ | soluble |
| (f) | lactic acid | $HC_3H_5O_3$ | soluble |

19.

| | Vitamin | Formula | Water or Fat Soluble |
|---|---|---|---|
| (a) | vitamin $B_1$ | $C_{12}H_{18}Cl_2N_4OS$ | water soluble |
| (b) | vitamin $B_3$ | $C_6H_6N_2O$ | water soluble |
| (c) | vitamin $B_6$ | $C_8H_{11}NO_3$ | water soluble |
| (d) | vitamin C | $C_6H_8O_6$ | water soluble |
| (e) | vitamin D | $C_{27}H_{44}O$ | fat soluble |
| (f) | vitamin K | $C_{31}H_{46}O_2$ | fat soluble |

## Section 14.4 *The Dissolving Process*

21. Fructose Dissolved in Water

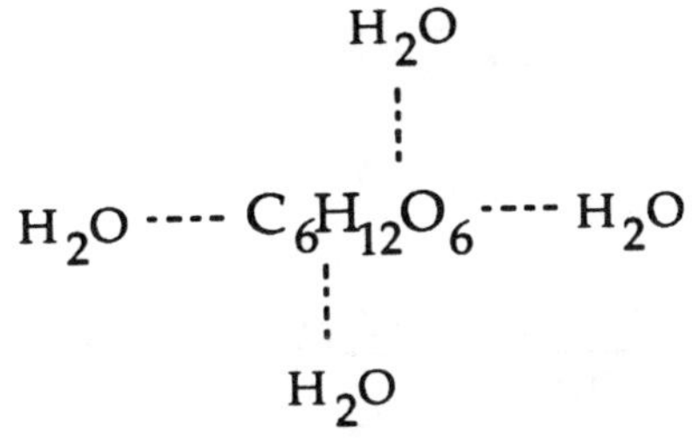

$H_2O$
⋮
$H_2O$ ---- $C_6H_{12}O_6$ ---- $H_2O$
⋮
$H_2O$

23. (a) Lithium Bromide Dissolved in Water

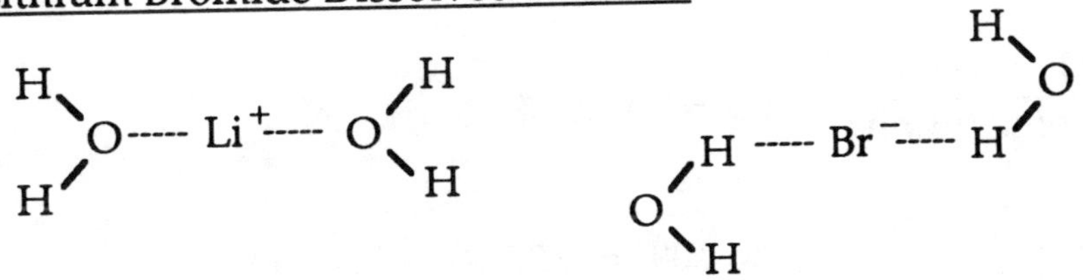

H\
O----- $Li^+$----- O
H/ (each O bonded to two H)

H
O---- H ---- $Br^-$----- H ---- O
H

(b) Calcium Chloride Dissolved in Water

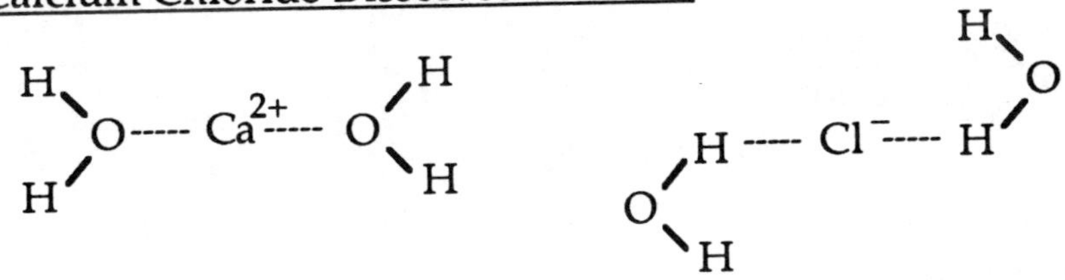

H
O----- $Ca^{2+}$----- O
H (each O bonded to two H)

H
O---- H ---- $Cl^-$----- H ---- O
H

**Section 14.5** *Rate of Dissolving*

25. Factors That Increase the Rate of Dissolving
    (1) heating the solution
    (2) stirring the solution
    (3) grinding the solid solute

27. (a) ~ 35 g $NaCl/100\ g\ H_2O$ (b) ~ 35 g $KCl/100\ g\ H_2O$

29. (a) ~ 36 g $NaCl/100\ g\ H_2O$ (b) ~ 41 g $KCl/100\ g\ H_2O$

31. (a) ~ 0°C (b) ~ 70°C

33. (a) ~ 100°C (b) ~ 50°C

**Section 14.7** *Unsaturated, Saturated, Supersaturated Solutions*

35. (a) supersaturated
    (b) saturated
    (c) unsaturated

37. (a) supersaturated
    (b) saturated
    (c) unsaturated

39. $$25.0\ \cancel{g\ H_2O} \times \frac{40.0\ g\ \text{rock salt}}{100\ \cancel{g\ H_2O}} = 10.0\ g\ \text{rock salt}$$

The solution is saturated.

41. | | Solute Remaining | | Solute that Crystallizes |
|---|---|---|---|
| (a) | ~ 80 g | (b) | ~ 20 g |

**Section 14.8** *Mass Percent Concentration*

43.* $$\frac{\text{mass of solute}}{\text{mass of solution}} \times 100 = m/m\ \%$$

(a) $$\frac{1.25\ g\ NaCl}{100.0\ g\ \text{solution}} \times 100 = 1.25\%$$

(b) $$\frac{2.50\ g\ K_2Cr_2O_7}{95.0\ g\ \text{solution}} \times 100 = 2.63\%$$

* continued on next page

43. (c) $\frac{10.0 \text{ g } CaCl_2}{250.0 \text{ g solution}} \times 100 = 4.00\%$

(d) $\frac{65.0 \text{ g sugar}}{125.0 \text{ g solution}} \times 100 = 52.0\%$

45.

| Solution | Unit Factors |
|---|---|
| (a) 1.50% KBr | $\frac{1.50 \text{ g KBr}}{100.00 \text{ g solution}}$ and $\frac{100.00 \text{ g solution}}{1.50 \text{ g KBr}}$ |
| | $\frac{98.50 \text{ g } H_2O}{100.00 \text{ g solution}}$ and $\frac{100.00 \text{ g solution}}{98.50 \text{ g } H_2O}$ |
| | $\frac{1.50 \text{ g KBr}}{98.50 \text{ g } H_2O}$ and $\frac{98.50 \text{ g } H_2O}{1.50 \text{ g KBr}}$ |
| (b) 2.50% $AlCl_3$ | $\frac{2.50 \text{ g } AlCl_3}{100.00 \text{ g solution}}$ and $\frac{100.00 \text{ g solution}}{2.50 \text{ g } AlCl_3}$ |
| | $\frac{97.50 \text{ g } H_2O}{100.00 \text{ g solution}}$ and $\frac{100.00 \text{ g solution}}{97.50 \text{ g } H_2O}$ |
| | $\frac{2.50 \text{ g } AlCl_3}{97.50 \text{ g } H_2O}$ and $\frac{97.50 \text{ g } H_2O}{2.50 \text{ g } AlCl_3}$ |
| (c) 3.75% $AgNO_3$ | $\frac{3.75 \text{ g } AgNO_3}{100.00 \text{ g solution}}$ and $\frac{100.00 \text{ g solution}}{3.75 \text{ g } AgNO_3}$ |
| | $\frac{96.25 \text{ g } H_2O}{100.00 \text{ g solution}}$ and $\frac{100.00 \text{ g solution}}{96.25 \text{ g } H_2O}$ |
| | $\frac{3.75 \text{ g } AgNO_3}{96.25 \text{ g } H_2O}$ and $\frac{96.25 \text{ g } H_2O}{3.75 \text{ g } AgNO_3}$ |
| (d) 4.25% $Li_2SO_4$ | $\frac{4.25 \text{ g } Li_2SO_4}{100.00 \text{ g solution}}$ and $\frac{100.00 \text{ g solution}}{4.25 \text{ g } Li_2SO_4}$ |
| | $\frac{95.75 \text{ g } H_2O}{100.00 \text{ g solution}}$ and $\frac{100.00 \text{ g solution}}{95.75 \text{ g } H_2O}$ |
| | $\frac{4.25 \text{ g } Li_2SO_4}{95.75 \text{ g } H_2O}$ and $\frac{95.75 \text{ g } H_2O}{4.25 \text{ g } Li_2SO_4}$ |

47. (a) $5.36\ \cancel{\text{g glucose}} \times \dfrac{100.0 \text{ g solution}}{10.0\ \cancel{\text{g glucose}}} = 53.6 \text{ g solution}$

(b) $25.0\ \cancel{\text{g sucrose}} \times \dfrac{100.0 \text{ g solution}}{12.5\ \cancel{\text{g sucrose}}} = 200 \text{ g solution } (2.00 \times 10^2 \text{ g})$

49. (a) $85.0\ \cancel{\text{g solution}} \times \dfrac{2.00 \text{ g } FeBr_2}{100.0\ \cancel{\text{g solution}}} = 1.70 \text{ g } FeBr_2$

(b) $105.0\ \cancel{\text{g solution}} \times \dfrac{5.00 \text{ g } Na_2CO_3}{100.0\ \cancel{\text{g solution}}} = 5.25 \text{ g } Na_2CO_3$

51. (a) $250.0\ \cancel{\text{g solution}} \times \dfrac{99.10 \text{ g } H_2O}{100.0\ \cancel{\text{g solution}}} = 247.8 \text{ g } H_2O$

(b) $100.0\ \cancel{\text{g solution}} \times \dfrac{95.00 \text{ g } H_2O}{100.0\ \cancel{\text{g solution}}} = 95.00 \text{ g } H_2O$

**Section 14.9** *Molar Concentration*

53. (a) MM of NaCl = 23.0 g + 35.5 g = 58.5 g/mol

100.0 mL solution = 0.1000 L solution

$$\frac{1.50\ \cancel{\text{g NaCl}}}{0.1000 \text{ L solution}} \times \frac{1 \text{ mol NaCl}}{58.5\ \cancel{\text{g NaCl}}} = 0.256\ M \text{ NaCl}$$

(b) MM of $K_2Cr_2O_7$ = 2(39.1) g + 2(52.0) g + 7(16.0) g = 294.2 g/mol

100.0 mL solution = 0.1000 L solution

$$\frac{1.50\ \cancel{\text{g } K_2Cr_2O_7}}{0.1000 \text{ L solution}} \times \frac{1 \text{ mol } K_2Cr_2O_7}{294.2\ \cancel{\text{g } K_2Cr_2O_7}} = 0.0510\ M\ K_2Cr_2O_7$$

(c) MM of $CaCl_2$ = 40.1 g + 2(35.5) g = 111.1 g/mol

125 mL solution = 0.125 L solution

$$\frac{5.55\ \cancel{\text{g } CaCl_2}}{0.125 \text{ L solution}} \times \frac{1 \text{ mol } CaCl_2}{111.1\ \cancel{\text{g } CaCl_2}} = 0.400\ M\ CaCl_2$$

(d) MM of $Na_2SO_4$ = 2(23.0) g + 32.1 g + 4(16.0) g = 142.1 g/mol

125 mL solution = 0.125 L solution

$$\frac{5.55\ \cancel{\text{g } Na_2SO_4}}{0.125 \text{ L solution}} \times \frac{1 \text{ mol } Na_2SO_4}{142.1\ \cancel{\text{g } Na_2SO_4}} = 0.312\ M\ Na_2SO_4$$

55. **Solution** **Unit Factors**

(a) 0.100 *M* LiI

$$\frac{0.100 \text{ mol LiI}}{1 \text{ L solution}} \text{ and } \frac{1 \text{ L solution}}{0.100 \text{ mol LiI}}$$

$$\frac{0.100 \text{ mol LiI}}{1000 \text{ mL solution}} \text{ and } \frac{1000 \text{ mL solution}}{0.100 \text{ mol LiI}}$$

(b) 0.100 *M* $NaNO_3$

$$\frac{0.100 \text{ mol NaNO}_3}{1 \text{ L solution}} \text{ and } \frac{1 \text{ L solution}}{0.100 \text{ mol NaNO}_3}$$

$$\frac{0.100 \text{ mol NaNO}_3}{1000 \text{ mL solution}} \text{ and } \frac{1000 \text{ mL solution}}{0.100 \text{ mol NaNO}_3}$$

(c) 0.500 *M* $K_2CrO_4$

$$\frac{0.500 \text{ mol K}_2\text{CrO}_4}{1 \text{ L solution}} \text{ and } \frac{1 \text{ L solution}}{0.500 \text{ mol K}_2\text{CrO}_4}$$

$$\frac{0.500 \text{ mol K}_2\text{CrO}_4}{1000 \text{ mL solution}} \text{ and } \frac{1000 \text{ mL solution}}{0.500 \text{ mol K}_2\text{CrO}_4}$$

(d) 0.500 *M* $ZnSO_4$

$$\frac{0.500 \text{ mol ZnSO}_4}{1 \text{ L solution}} \text{ and } \frac{1 \text{ L solution}}{0.500 \text{ mol ZnSO}_4}$$

$$\frac{0.500 \text{ mol ZnSO}_4}{1000 \text{ mL solution}} \text{ and } \frac{1000 \text{ mL solution}}{0.500 \text{ mol ZnSO}_4}$$

57.* (a) MM of NaF = 23.0 g + 19.0 g = 42.0 g/mol

$$10.0 \cancel{\text{ g NaF}} \times \frac{1 \cancel{\text{ mol NaF}}}{42.0 \cancel{\text{ g NaF}}} \times \frac{1 \text{ L solution}}{0.275 \cancel{\text{ mol NaF}}} = 0.866 \text{ L solution}$$

(b) MM of $CdCl_2$ = 112.4 g + 2(35.5) g = 183.4 g/mol

$$10.0 \cancel{\text{ g CdCl}_2} \times \frac{1 \cancel{\text{ mol CdCl}_2}}{183.4 \cancel{\text{ g CdCl}_2}} \times \frac{1 \text{ L solution}}{0.275 \cancel{\text{ mol CdCl}_2}} = 0.198 \text{ L solution}$$

(c) MM of $K_2CO_3$ = 2(39.1) g + 12.0 g + 3(16.0) = 138.2 g/mol

$$10.0 \cancel{\text{ g K}_2\text{CO}_3} \times \frac{1 \cancel{\text{ mol K}_2\text{CO}_3}}{138.2 \cancel{\text{ g K}_2\text{CO}_3}} \times \frac{1 \text{ L solution}}{0.408 \cancel{\text{ mol K}_2\text{CO}_3}}$$

$$= 0.177 \text{ L solution}$$

* continued on next page

57. (d) MM of $Fe(ClO_3)_3$ = 55.8 g + 3(35.5) g + 9(16.0) g = 306.3 g/mol

$$10.0\ g\ \cancel{Fe(ClO_3)_3} \times \frac{1\ \cancel{mol\ Fe(ClO_3)_3}}{306.3\ g\ \cancel{Fe(ClO_3)_3}} \times \frac{1\ L\ solution}{0.408\ \cancel{mol\ Fe(ClO_3)_3}} = 0.0800\ L\ solution$$

59. (a) MM of NaOH = 23.0 g + 16.0 g + 1.0 g = 40.0 g/mol

$$1.00\ \cancel{L\ solution} \times \frac{0.100\ \cancel{mol\ NaOH}}{1\ \cancel{L\ solution}} \times \frac{40.0\ g\ NaOH}{1\ \cancel{mol\ NaOH}} = 4.00\ g\ NaOH$$

(b) MM of $LiHCO_3$ = 6.9 g + 1.0 g + 12.0 g + 3(16.0) g = 67.9 g/mol

$$1.00\ \cancel{L\ solution} \times \frac{0.100\ \cancel{mol\ LiHCO_3}}{1\ \cancel{L\ solution}} \times \frac{67.9\ g\ LiHCO_3}{1\ \cancel{mol\ LiHCO_3}} = 6.79\ g\ LiHCO_3$$

(c) MM of $CuCl_2$ = 63.5 g + 2(35.5) g = 134.5 g/mol

$$25.0\ \cancel{mL\ solution} \times \frac{0.500\ \cancel{mol\ CuCl_2}}{1000\ \cancel{mL\ solution}} \times \frac{134.5\ g\ CuCl_2}{1\ \cancel{mol\ CuCl_2}} = 1.68\ g\ CuCl_2$$

(d) MM of $KMnO_4$ = 39.1 g + 54.9 g + 4(16.0) g = 158.0g/mol

$$25.0\ \cancel{mL\ solution} \times \frac{0.500\ \cancel{mol\ KMnO_4}}{1000\ \cancel{mL\ solution}} \times \frac{158.0\ g\ KMnO_4}{1\ \cancel{mol\ KMnO_4}} = 1.98\ g\ KMnO_4$$

61. MM of $CaSO_4$ = 40.1 g + 32.1 g + 4(16.0) g = 136.2 g/mol

100 mL solution = 0.100 L solution

$$\frac{0.209\ \cancel{g\ CaSO_4}}{0.100\ L\ solution} \times \frac{1\ mol\ CaSO_4}{136.2\ \cancel{g\ CaSO_4}} = 0.0153\ M\ CaSO_4$$

63. (a) $\frac{0.524\ g\ glucose}{10.483\ g\ solution} \times 100 = 5.00\%$

(b) MM of $C_6H_{12}O_6$ = 6(12.0) g + 12(1.0) g + 6(16.0) g = 180.0 g/mol

10.0 mL solution = 0.0100 L solution

$$\frac{0.524\ \cancel{g\ C_6H_{12}O_6}}{0.0100\ L\ solution} \times \frac{1\ mol\ C_6H_{12}O_6}{180.0\ \cancel{g\ C_6H_{12}O_6}} = 0.291\ M\ C_6H_{12}O_6$$

**Section 14.10** *Molal Concentration and Colligative Properties*

65. (a) MM of KF = 39.1 g + 19.0 g = 58.1 g/mol

$$\frac{10.0\ \cancel{g\ KF}}{2.50\ kg\ H_2O} \times \frac{1\ mol\ KF}{58.1\ \cancel{g\ KF}} = 0.0688\ m\ KF$$

(b) MM of $ZnSO_4$ = 65.4 g + 32.1 g + 4(16.0) g = 161.5 g/mol

375 g $H_2O$ = 0.375 kg $H_2O$

$$\frac{10.0\ \cancel{g\ ZnSO_4}}{0.375\ kg\ H_2O} \times \frac{1\ mol\ ZnSO_4}{161.5\ \cancel{g\ ZnSO_4}} = 0.165\ m\ ZnSO_4$$

67. MM of $C_{12}H_{22}O_{11}$ = 12(12.0) g + 22(1.0) g + 11(16.0) g = 342.0 g/mol

$$6.5\ \cancel{kg\ H_2O} \times \frac{2.00\ \cancel{mol\ C_{12}H_{22}O_{11}}}{1\ \cancel{kg\ H_2O}} \times \frac{342.0\ g\ C_{12}H_{22}O_{11}}{1\ \cancel{mol\ C_{12}H_{22}O_{11}}}$$
$$= 4400\ g\ C_{12}H_{22}O_{11}$$

69. $m\ K_f = \Delta T_f$

500.0 g $H_2O$ = 0.5000 kg $H_2O$

$$m = \frac{100.0\ \cancel{g\ CH_3OH}}{0.5000\ kg\ H_2O} \times \frac{1\ mol\ CH_3OH}{32.0\ \cancel{g\ CH_3OH}} = 6.25\ m$$

$$K_f = \frac{1.86°C}{m}$$

$$\Delta T_f = 6.25\ \cancel{m} \times \frac{1.86°C}{\cancel{m}} = 11.6°C$$

Freezing point of solution: – 11.6°C

71. $$m = \frac{\Delta T_f}{K_f}$$

$$\Delta T_f = [0.00°C - (-3.72°C)] = 3.72°C$$

$$K_f = \frac{1.86°C}{m}$$

$$m = 3.72\cancel{°C} \times \frac{m}{1.86\cancel{°C}} = 2.00\ m$$

100.0 g $H_2O$ = 0.1000 kg $H_2O$

$$\text{MM of unknown} = \frac{36.0\ g}{0.1000\ \cancel{kg\ H_2O}} \times \frac{1\ \cancel{kg\ H_2O}}{2.00\ mol} = 180\ g/mol$$

73. $m = \frac{\Delta T_f}{K_f}$

$\Delta T_f = [-117.3°C - (-119.8°C)] = 2.5°C$

$K_f = \frac{1.99°C}{m}$

$m = 2.5\cancel{°C} \times \frac{m}{1.99\cancel{°C}} = 1.3\ m$

100.0 g ethyl alcohol = 0.1000 kg ethyl alcohol

$\text{MM of unknown} = \frac{4.50\text{ g}}{0.1000\ \cancel{\text{kg ethyl alcohol}}} \times \frac{1\ \cancel{\text{kg ethyl alcohol}}}{1.3\text{ mol}}$

$= 35\text{ g/mol}$

**Section 14.11** *Colloids*

75. The water droplets in the Tyndall effect serve as a colloidal dispersion. Thus, the water droplets must range in size from 1 to 100 nanometers.

77.

| | Observation | Solution or Colloid |
|---|---|---|
| (a) | dispersed particles separate in a centrifuge | colloid |
| (b) | dispersed particles demonstrate the Tyndall effect | colloid |
| (c) | dispersed particles pass through a semipermeable membrane | solution |

**General Exercises**

79. $\frac{0.0019\text{ g }N_2}{100\text{ g blood}} \times \frac{4.68\ \cancel{\text{atm}}}{1.00\ \cancel{\text{atm}}} = 0.0089\text{ g }N_2/100\text{ g blood}$

81. 100 mL solution = 0.100 L solution

$1.00\ \cancel{\text{L solution}} \times \frac{22.8\text{ g }SO_2}{0.100\ \cancel{\text{L solution}}} = 228\text{ g }SO_2$

83. Air is less soluble in hot water than in cold tap water. Therefore, air leaves the solution and forms bubbles on the sides of the pan.

85. The polar –OH on an alcohol can hydrogen bond with water causing it to be soluble. As the nonpolar $C_xH_y$– portion of the molecule increases in size, the overall molecule becomes less polar and eventually immiscible with water.

87.

| | Solute | Solvent |
|---|---|---|
| (a) | ethyl alcohol | water |
| (b) | water | ethyl alcohol |

89. MM of NaClO = 23.1 g + 35.5 g + 16.0 g = 74.5 g/mol

Solute: $5.25 \text{ }\cancel{\text{g NaClO}} \times \dfrac{1 \text{ mol NaClO}}{74.5 \text{ }\cancel{\text{g NaClO}}} = 0.0705 \text{ mol NaClO}$

Solution: $100 \text{ }\cancel{\text{g bleach}} \times \dfrac{1 \text{ }\cancel{\text{mL bleach}}}{1.04 \text{ }\cancel{\text{g bleach}}} \times \dfrac{1 \text{ L bleach}}{1000 \text{ }\cancel{\text{mL bleach}}}$
$= 0.0962 \text{ L bleach}$

Molarity: $\dfrac{0.0705 \text{ mol NaClO}}{0.0962 \text{ L bleach}} = 0.733\ M \text{ NaClO}$

91. $\dfrac{0.0075 \text{ }\cancel{\text{g glucose}}}{10.0 \text{ }\cancel{\text{mL blood}}} \times \dfrac{1000 \text{ mg glucose}}{1 \text{ }\cancel{\text{g glucose}}} \times \dfrac{1000 \text{ }\cancel{\text{mL blood}}}{1 \text{ }\cancel{\text{L blood}}} \times \dfrac{1 \text{ }\cancel{\text{L blood}}}{10 \text{ dL blood}}$
$= 75 \text{ mg glucose/dL blood}$

Since the calculated value is between 70 to 90 milligrams per deciliter, the blood glucose reading is normal.

# CHAPTER 15

# Acids and Bases

## Section 15.1 *Properties of Acids and Bases*

1. General Properties of Acids
   (1) Acids have a sour taste
   (2) Acids have a pH < 7
   (3) Acids turn blue litmus paper red

3.

| | Food | pH | Classification |
|---|---|---|---|
| (a) | egg white | 7.9 | basic |
| (b) | sour milk | 6.2 | acidic |
| (c) | maple syrup | 7.0 | neutral |
| (d) | lime juice | 1.8 | acidic |
| (e) | champagne | 3.8 | acidic |
| (f) | tomato juice | 4.1 | acidic |

## Section 15.2 *Acid-Base Indicators*

5.

| | pH | Methyl Red Color | | pH | Methyl Red Color |
|---|---|---|---|---|---|
| (a) | 3 | red | (b) | 7 | yellow |

7.

| | pH | Phenolphthalein Color | | pH | Phenolphthalein Color |
|---|---|---|---|---|---|
| (a) | 7 | colorless | (b) | 11 | pink |

9. At a pH of 7 the bromthymol blue indicator is colored green.

## Section 15.3 *Standard Solutions of Acids and Bases*

11.* $2\ HNO_{3(aq)} + Na_2CO_{3(s)} \rightarrow 2\ NaNO_{3(aq)} + H_2O_{(l)} + CO_{2(g)}$

MM of $Na_2CO_3$ = 106.0 g/mol

$$0.689\ \cancel{g\ Na_2CO_3} \times \frac{1\ \cancel{mol\ Na_2CO_3}}{106.0\ \cancel{g\ Na_2CO_3}} \times \frac{2\ mol\ HNO_3}{1\ \cancel{mol\ Na_2CO_3}} = 0.0130\ mol\ HNO_3$$

* continued on next page

11. $$\frac{0.0130 \text{ mol } HNO_3}{41.25 \cancel{\text{ mL solution}}} \times \frac{1000 \cancel{\text{ mL solution}}}{1 \text{ L solution}} = \frac{0.315 \text{ mol } HNO_3}{1 \text{ L solution}}$$

Molarity of nitric acid: 0.315 *M* $HNO_3$

13. $2\ HCl_{(aq)} + Na_2C_2O_{4(s)} \rightarrow H_2C_2O_{4(aq)} + 2\ NaCl_{(aq)}$

MM of $Na_2C_2O_4$ = 134.0 g/mol

$$1.550 \cancel{\text{ g } Na_2C_2O_4} \times \frac{1 \cancel{\text{ mol } Na_2C_2O_4}}{134.0 \cancel{\text{ g } Na_2C_2O_4}} \times \frac{2 \text{ mol HCl}}{1 \cancel{\text{ mol } Na_2C_2O_4}} = 0.02313 \text{ mol HCl}$$

$$\frac{0.02313 \text{ mol HCl}}{20.95 \cancel{\text{ mL solution}}} \times \frac{1000 \cancel{\text{ mL solution}}}{1 \text{ L solution}} = \frac{1.104 \text{ mol HCl}}{1 \text{ L solution}}$$

Molarity of hydrochloric acid: 1.104 *M* HCl

15. $H_2C_2O_{4(s)} + 2\ LiOH_{(aq)} \rightarrow Li_2C_2O_{4(aq)} + 2\ H_2O_{(l)}$

MM of $H_2C_2O_4$ = 90.0 g/mol

$$0.627 \cancel{\text{ g } H_2C_2O_4} \times \frac{1 \cancel{\text{ mol } H_2C_2O_4}}{90.0 \cancel{\text{ g } H_2C_2O_4}} \times \frac{2 \text{ mol LiOH}}{1 \cancel{\text{ mol } H_2C_2O_4}} = 0.0139 \text{ mol LiOH}$$

$$\frac{0.0139 \text{ mol LiOH}}{29.00 \cancel{\text{ mL solution}}} \times \frac{1000 \cancel{\text{ mL solution}}}{1 \text{ L solution}} = \frac{0.479 \text{ mol LiOH}}{1 \text{ L solution}}$$

Molarity of lithium hydroxide: 0.479 *M* LiOH

17. $HAsc_{(s)} + NaOH_{(aq)} \rightarrow NaAsc_{(aq)} + H_2O_{(l)}$

$$30.95 \cancel{\text{ mL solution}} \times \frac{0.176 \text{ mol NaOH}}{1000 \cancel{\text{ mL solution}}} = 0.00545 \text{ mol NaOH}$$

$$0.00545 \cancel{\text{ mol NaOH}} \times \frac{1 \text{ mol HAsc}}{1 \cancel{\text{ mol NaOH}}} = 0.00545 \text{ mol HAsc}$$

$$\text{MM of vitamin C} = \frac{0.959 \text{ g HAsc}}{0.00545 \text{ mol HAsc}} = 176 \text{ g/mol}$$

19. Alanine is abbreviated HAla

$HAla_{(s)} + NaOH_{(aq)} \rightarrow NaAla_{(aq)} + H_2O_{(l)}$

$$21.05 \cancel{\text{mL solution}} \times \frac{0.145 \text{ mol NaOH}}{1000 \cancel{\text{mL solution}}} = 0.00305 \text{ mol NaOH}$$

$$0.00305 \cancel{\text{mol NaOH}} \times \frac{1 \text{ mol HAla}}{1 \cancel{\text{mol NaOH}}} = 0.00305 \text{ mol HAla}$$

$$\text{MM of alanine} = \frac{0.272 \text{ g HAla}}{0.00305 \text{ mol HAla}} = 89.2 \text{ g/mol}$$

**Section 15.4** *Acid-Base Titrations*

21. $HCl_{(aq)} + NaOH_{(aq)} \rightarrow NaCl_{(aq)} + H_2O_{(l)}$

$$22.15 \cancel{\text{mL solution}} \times \frac{0.155 \text{ mol NaOH}}{1000 \cancel{\text{mL solution}}} = 0.00343 \text{ mol NaOH}$$

$$0.00343 \cancel{\text{mol NaOH}} \times \frac{1 \text{ mol HCl}}{1 \cancel{\text{mol NaOH}}} = 0.00343 \text{ mol HCl}$$

$$\frac{0.00343 \text{ mol HCl}}{25.0 \cancel{\text{mL solution}}} \times \frac{1000 \cancel{\text{mL solution}}}{1 \text{ L solution}} = \frac{0.137 \text{ mol HCl}}{1 \text{ L solution}}$$

Molarity of hydrochloric acid: 0.137 *M* HCl

23. $H_3PO_{4(aq)} + 3\ NaOH_{(aq)} \rightarrow Na_3PO_{4(aq)} + 3\ H_2O_{(l)}$

$$34.45 \cancel{\text{mL solution}} \times \frac{0.210 \text{ mol NaOH}}{1000 \cancel{\text{mL solution}}} = 0.00723 \text{ mol NaOH}$$

$$0.00723 \cancel{\text{mol NaOH}} \times \frac{1 \text{ mol } H_3PO_4}{3 \cancel{\text{mol NaOH}}} = 0.00241 \text{ mol } H_3PO_4$$

$$\frac{0.00241 \text{ mol } H_3PO_4}{28.55 \cancel{\text{mL solution}}} \times \frac{1000 \cancel{\text{mL solution}}}{1 \text{ L solution}} = \frac{0.0844 \text{ mol } H_3PO_4}{1 \text{ L solution}}$$

Molarity of phosphoric acid: 0.0844 *M* $H_3PO_4$

25. $H_2SO_{4(aq)} + 2\ KOH_{(aq)} \rightarrow K_2SO_{4(aq)} + 2\ H_2O_{(l)}$

$$41.05\ \cancel{\text{mL solution}} \times \frac{0.165\ \text{mol KOH}}{1000\ \cancel{\text{mL solution}}} = 0.00677\ \text{mol KOH}$$

$$0.00677\ \cancel{\text{mol KOH}} \times \frac{1\ \text{mol}\ H_2SO_4}{2\ \cancel{\text{mol KOH}}} = 0.00339\ \text{mol}\ H_2SO_4$$

$$0.00339\ \cancel{\text{mol}\ H_2SO_4} \times \frac{1000\ \text{mL solution}}{0.122\ \cancel{\text{mol}\ H_2SO_4}} = 27.8\ \text{mL}\ H_2SO_4$$

27. (a) MM of HCl = 36.5 g/mol

$$\frac{6.00\ \cancel{\text{mol HCl}}}{1000\ \cancel{\text{mL solution}}} \times \frac{36.5\ \text{g HCl}}{1\ \cancel{\text{mol HCl}}} \times \frac{1\ \cancel{\text{mL solution}}}{1.10\ \text{g solution}} \times 100$$
$$= 19.9\%\ \text{HCl}$$

(b) MM of $HC_2H_3O_2$ = 60.0 g/mol

$$\frac{1.00\ \cancel{\text{mol}\ HC_2H_3O_2}}{1000\ \cancel{\text{mL solution}}} \times \frac{60.0\ \text{g}\ HC_2H_3O_2}{1\ \cancel{\text{mol}\ HC_2H_3O_2}} \times \frac{1\ \cancel{\text{mL solution}}}{1.01\ \text{g solution}} \times 100$$
$$= 5.94\%\ HC_2H_3O_2$$

(c) MM of $HNO_3$ = 63.0 g/mol

$$\frac{0.500\ \cancel{\text{mol}\ HNO_3}}{1000\ \cancel{\text{mL solution}}} \times \frac{63.0\ \text{g}\ HNO_3}{1\ \cancel{\text{mol}\ HNO_3}} \times \frac{1\ \cancel{\text{mL solution}}}{1.01\ \text{g solution}} \times 100$$
$$= 3.12\%\ HNO_3$$

(d) MM of $H_2SO_4$ = 98.1 g/mol

$$\frac{3.00\ \cancel{\text{mol}\ H_2SO_4}}{1000\ \cancel{\text{mL solution}}} \times \frac{98.1\ \text{g}\ H_2SO_4}{1\ \cancel{\text{mol}\ H_2SO_4}} \times \frac{1\ \cancel{\text{mL solution}}}{1.18\ \text{g solution}} \times 100$$
$$= 24.9\%\ H_2SO_4$$

(e) MM of $H_3PO_4$ = 98.0 g/mol

$$\frac{0.631\ \cancel{\text{mol}\ H_3PO_4}}{1000\ \cancel{\text{mL solution}}} \times \frac{98.0\ \text{g}\ H_3PO_4}{1\ \cancel{\text{mol}\ H_3PO_4}} \times \frac{1\ \cancel{\text{mL solution}}}{1.03\ \text{g solution}} \times 100$$
$$= 6.00\%\ H_3PO_4$$

## Section 15.5 *Arrhenius Acid-Base Theory*

29.

| | Acid | Degree of Ionization | Strength |
|---|---|---|---|
| (a) | $HClO_{3(aq)}$ | ~ 100% | strong |
| (b) | $HIO_{(aq)}$ | ~ 1% | weak |
| (c) | $HBr_{(aq)}$ | ~ 100% | strong |
| (d) | $HC_7H_5O_{2(aq)}$ | ~ 1% | weak |

31.

| | Formula | Classification | | Formula | Classification |
|---|---|---|---|---|---|
| (a) | $HClO_{(aq)}$ | Arrhenius acid | (b) | $KOH_{(aq)}$ | Arrhenius base |
| (c) | $K_2SO_{4(aq)}$ | salt | (d) | $Ba(OH)_{2(aq)}$ | Arrhenius base |
| (e) | $HNO_{3(aq)}$ | Arrhenius acid | (f) | $Mg(NO_3)_{2(aq)}$ | salt |
| (g) | $Ca(OH)_{2(aq)}$ | Arrhenius base | (h) | $H_2SO_{3(aq)}$ | Arrhenius acid |

33.

| | Salt | Acid | Base |
|---|---|---|---|
| (a) | $KBr_{(aq)}$ | $HBr$ | $KOH$ |
| (b) | $BaCl_{2(aq)}$ | $HCl$ | $Ba(OH)_2$ |
| (c) | $Ca(NO_3)_{2(aq)}$ | $HNO_3$ | $Ca(OH)_2$ |
| (d) | $Na_3PO_{4(aq)}$ | $H_3PO_4$ | $NaOH$ |
| (e) | $CoSO_{4(aq)}$ | $H_2SO_4$ | $Co(OH)_2$ |
| (f) | $Li_2CO_{3(aq)}$ | $H_2CO_3$ | $LiOH$ |

**Section 15.6** *Brønsted-Lowry Acid-Base Theory*

35.

| | Acid | Base | | Acid | Base |
|---|---|---|---|---|---|
| (a) | $HC_2H_3O_{2(aq)}$ | $LiOH_{(aq)}$ | (b) | $HBr_{(aq)}$ | $NaCN_{(aq)}$ |
| (c) | $HClO_{4(aq)}$ | $K_2CO_{3(aq)}$ | (d) | $H_2SO_{4(aq)}$ | $NH_{3(aq)}$ |

37.

| | Stronger Acid | Stronger Base | | Stronger Acid | Stronger Base |
|---|---|---|---|---|---|
| (a) | $H_2SO_{3(aq)}$ | $NaHS_{(aq)}$ | (b) | $NaHSO_{4(aq)}$ | $NaF_{(aq)}$ |
| (c) | $H_2PO_4^-{}_{(aq)}$ | $NH_{3(aq)}$ | (d) | $H_2SO_{4(aq)}$ | $H_2O_{(aq)}$ |

**Section 15.7** *Ionization of Water*

39. (a) Equilibrium ionization reaction: $H_2O_{(l)} \rightleftarrows H^+_{(aq)} + OH^-_{(aq)}$
(b) Equilibrium constant expression: $K_w = [H^+]\,[OH^-]$
(c) Ionization constant for water at 25°C: $K_w = 1.0 \times 10^{-14}$
(d) Molar hydrogen ion concentration at 25°C: $[H^+] = 1.0 \times 10^{-7}$
(e) Molar hydroxide ion concentration at 25°C: $[OH^-] = 1.0 \times 10^{-7}$

41.

| | Change | $[H^+]$ | | Change | $[H^+]$ |
|---|---|---|---|---|---|
| (a) | increase $[OH^-]$ | decreases | (b) | decreases $[OH^-]$ | increases |

**Section 15.8** *The pH Concept*

43.* $pH = -\log [H^+]$
(a) $pH = -\log 0.001 = -\log 10^{-3}$
$pH = -(-3) = 3$
(b) $pH = -\log 0.000\,01 = -\log 10^{-5}$
$pH = -(-5) = 5$

---

* continued on next page

43. (c) $pH = -\log 0.000\,000\,01 = -\log 10^{-8}$
$pH = -(-8) = 8$
(d) $pH = -\log 0.000\,001 = -\log 10^{-6}$
$pH = -(-6) = 6$

**Section 15.9** *Advanced pH and pOH Calculations*

45. (a) $pH = -\log 0.000\,0079 = -\log 7.9 \times 10^{-6}$
$pH = -\log 7.9 - \log 10^{-6}$
$pH = -0.90 - (-6) = 5.10$
(b) $pH = -\log 0.000\,000\,39 = -\log 3.9 \times 10^{-7}$
$pH = -\log 3.9 - \log 10^{-7}$
$pH = -0.59 - (-7) = 6.41$
(c) $pH = -\log 0.000\,000\,30 = -\log 3.0 \times 10^{-7}$
$pH = -\log 3.0 - \log 10^{-7}$
$pH = -0.48 - (-7) = 6.52$
(d) $pH = -\log 0.000\,000\,016 = -\log 1.6 \times 10^{-8}$
$pH = -\log 1.6 - \log 10^{-8}$
$pH = -0.20 - (-8) = 7.80$

47. (a) $pOH = -\log 0.11 = -\log 1.1 \times 10^{-1}$
$pOH = -\log 1.1 - \log 10^{-1}$
$pOH = -0.04 - (-1) = 0.96$
(b) $pOH = -\log 0.000\,55 = -\log 5.5 \times 10^{-4}$
$pOH = -\log 5.5 - \log 10^{-4}$
$pOH = -0.74 - (-4) = 3.26$
(c) $pOH = -\log 0.000\,31 = -\log 3.1 \times 10^{-4}$
$pOH = -\log 3.1 - \log 10^{-4}$
$pOH = -0.49 - (-4) = 3.51$
(d) $pOH = -\log 0.000\,000\,000\,66 = -\log 6.6 \times 10^{-10}$
$pOH = -\log 6.6 - \log 10^{-10}$
$pOH = -0.82 - (-10) = 9.18$

49. (a) $pOH = 14.00 - 0.30 = 13.70$
(b) $pOH = 14.00 - 2.52 = 11.48$

**Section 15.10** *Strong and Weak Electrolytes*

51.*

| | Substance | Conductivity | Classification |
|---|---|---|---|
| (a) | $HBrO_{(aq)}$ | weak | weak acid |
| (b) | $Sr(NO_3)_{2(aq)}$ | strong | soluble ionic compound |
| (c) | $HI_{(aq)}$ | strong | strong acid |
| (d) | $HC_2H_3O_{2(aq)}$ | weak | weak acid |

* continued on next page

51.

| | Substance | Conductivity | Classification |
|---|---|---|---|
| (e) | $KOH_{(aq)}$ | strong | strong base |
| (f) | $Sr(OH)_{2(aq)}$ | strong | strong base |
| (g) | $HClO_{4(aq)}$ | strong | strong acid |
| (h) | $NH_4OH_{(aq)}$ | weak | weak base |
| (i) | $KC_2H_3O_{2(aq)}$ | strong | soluble ionic compound |
| (j) | $PbI_{2(s)}$ | weak | very slightly soluble compound |

53.

| | Substance | Electrolyte | In Aqueous Solution |
|---|---|---|---|
| (a) | $HClO_{3(aq)}$ | strong | $H^+_{(aq)}$ and $ClO_3^-{}_{(aq)}$ |
| (b) | $Ca(NO_3)_{2(aq)}$ | strong | $Ca^{2+}_{(aq)}$ and 2 $NO_3^-{}_{(aq)}$ |
| (c) | $HBr_{(aq)}$ | strong | $H^+_{(aq)}$ and $Br^-_{(aq)}$ |
| (d) | $HClO_{2(aq)}$ | weak | $HClO_{2(aq)}$ |
| (e) | $LiOH_{(aq)}$ | strong | $Li^+_{(aq)}$ and $OH^-_{(aq)}$ |
| (f) | $Al(OH)_{3(s)}$ | weak | $Al(OH)_{3(s)}$ |
| (g) | $HNO_{2(aq)}$ | weak | $HNO_{2(aq)}$ |
| (h) | $HCHO_{2(aq)}$ | weak | $HCHO_{2(aq)}$ |
| (i) | $Cd(C_2H_3O_2)_{2(aq)}$ | strong | $Cd^{2+}_{(aq)}$ and 2 $C_2H_3O_2^-{}_{(aq)}$ |
| (j) | $K_2SO_{4(aq)}$ | strong | 2 $K^+_{(aq)}$ and $SO_4^{2-}{}_{(aq)}$ |

**Section 15.11** *Net Ionic Equations*

55. (1) Complete and balance the nonionized (molecular) equation.
(2) Convert the nonionized equation to the total ionic equation.
(3) Cancel spectator ions to obtain the net ionic equation.
(4) Check (√) each ion or atom on both sides of the equation.

57.* (a) molecular equation:

$HCl_{(aq)} + KOH_{(aq)} \rightarrow KCl_{(aq)} + H_2O_{(l)}$

total ionic equation:

$H^+_{(aq)} + \cancel{Cl}^-_{(aq)} + \cancel{K}^+_{(aq)} + OH^-_{(aq)} \rightarrow \cancel{K}^+_{(aq)} + \cancel{Cl}^-_{(aq)} + H_2O_{(l)}$

net ionic equation:

$H^+_{(aq)} + OH^-_{(aq)} \rightarrow H_2O_{(l)}$

(b) molecular equation:

$2\ HC_2H_3O_{2(aq)} + Ca(OH)_{2(aq)} \rightarrow Ca(C_2H_3O_2)_{2(aq)} + 2\ H_2O_{(l)}$

total ionic equation:

$2\ HC_2H_3O_{2(aq)} + \cancel{Ca}^{2+}_{(aq)} + 2\ OH^-_{(aq)} \rightarrow$
$\cancel{Ca}^{2+}_{(aq)} + 2\ C_2H_3O_2^-{}_{(aq)} + 2\ H_2O_{(l)}$

* continued on next page

57. (b) net ionic equation:

$HC_2H_3O_2(aq) + OH^-(aq) \rightarrow C_2H_3O_2^-(aq) + H_2O(l)$

(c) molecular equation:

$2\ HF(aq) + Li_2CO_3(aq) \rightarrow 2\ LiF(aq) + H_2O(l) + CO_2(g)$

total ionic equation:

$2\ HF(aq) + \cancel{2\ Li}^+(aq) + CO_3^{2-}(aq) \rightarrow \cancel{2\ Li}^+(aq) + 2\ F^-(aq) + H_2O(l) + CO_2(g)$

net ionic equation:

$2\ HF(aq) + CO_3^{2-}(aq) \rightarrow 2\ F^-(aq) + H_2O(l) + CO_2(g)$

(d) molecular equation:

$HNO_3(aq) + KHCO_3(aq) \rightarrow KNO_3(aq) + H_2O(l) + CO_2(g)$

total ionic equation:

$H^+(aq) + \cancel{NO_3}^-(aq) + \cancel{K}^+(aq) + HCO_3^-(aq) \rightarrow$
$\cancel{K}^+(aq) + \cancel{NO_3}^-(aq) + H_2O(l) + CO_2(g)$

net ionic equation:

$H^+(aq) + HCO_3^-(aq) \rightarrow H_2O(l) + CO_2(g)$

(e) molecular equation:

$H_2SO_4(aq) + Ba(OH)_2(aq) \rightarrow BaSO_4(s) + 2\ H_2O(l)$

total ionic equation:

$2\ H^+(aq) + SO_4^{2-}(aq) + Ba^{2+}(aq) + 2\ OH^-(aq) \rightarrow BaSO_4(s) + 2\ H_2O(l)$

net ionic equation:

$H^+(aq) + SO_4^{2-}(aq) + Ba^{2+}(aq) + OH^-(aq) \rightarrow BaSO_4(s) + H_2O(l)$

## General Exercises

59. A pH of 3.8 is in the middle of the color change range (3.2 to 4.4) for methyl orange. Thus, the indicator will contain both red and yellow components which gives an orange color.

61. The water (pH = 7) is yellow because methyl red is yellow at a pH of 7 while phenolphthalein is colorless at a pH of 7.

63. The species $H_2PO_4^-(aq)$ is a proton donor in the first reaction and a proton acceptor in the second reaction. Thus, $H_2PO_4^-(aq)$ is amphiprotic.

65. $1 \cancel{\text{mL water}} \times \frac{3 \times 10^{22} \cancel{\text{molecules water}}}{1 \cancel{\text{mL water}}} \times \frac{2 \text{ ions}}{1 \times 10^{9} \cancel{\text{molecules water}}}$

$= 6 \times 10^{13}$ water molecules existing as ions

67. (a) If the pH increases, the $[OH^-]$ increases.
(b) If the pOH increases, the $[H^+]$ increases.

69. A weak electrolyte solution conducts electricity slightly while a nonelectrolyte solution does not conduct electricity at all.

# Chemical Equilibrium

CHAPTER 16

**Section 16.1** *Collision Theory of Reaction Rates*

1. Factors That Influence Rates of Reaction
   (1) frequency of collisions
   (2) energy of collisions
   (3) orientation of colliding molecules (collision geometry)

3. Effective Collision Geometry

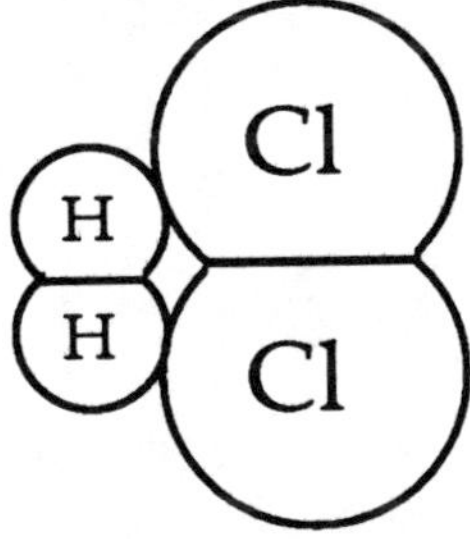


5.

| | Change | Effect on the Rate of Reaction |
|---|---|---|
| (a) | increase the concentration of a reactant | The rate of reaction increases because more collisions occur. |
| (b) | decrease the temperature of the reaction | The rate of reaction decreases because both collision frequency and collision energy decrease. |
| (c) | add a catalyst | The rate of reaction increases because collision geometry is made more effective. |

7. In a mine, fine particles of coal dust in the air provide a greater surface area for reaction. In the barbecue, large briquettes have less surface area to contact the oxygen in air.

9. $PCl_{5(g)}$ + heat → $PCl_{3(g)} + Cl_{2(g)}$

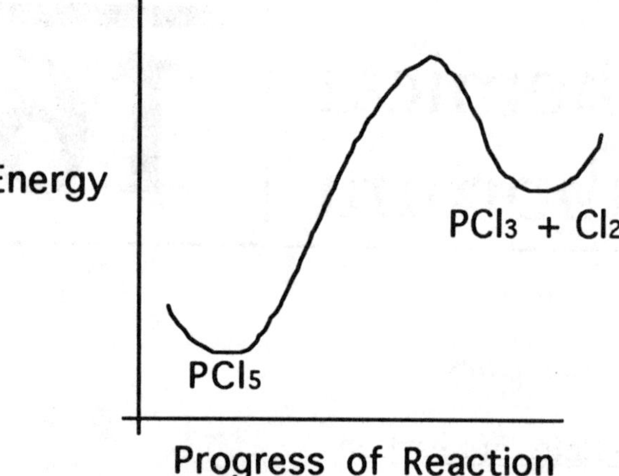


11. $H_{2(g)} + I_{2(g)}$ → 2 $HI_{(g)}$

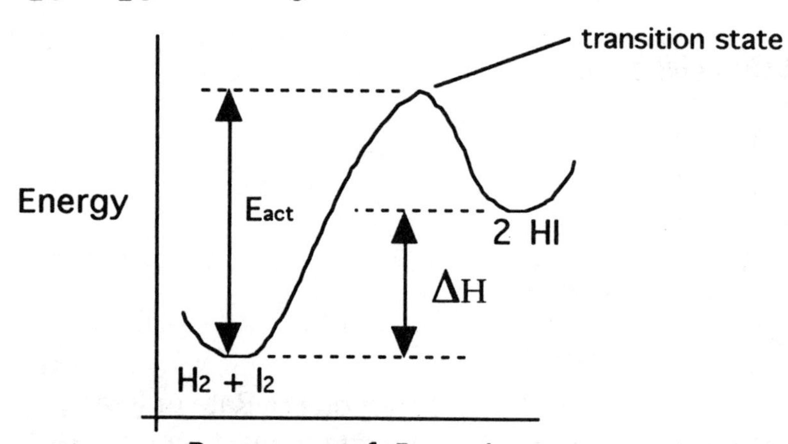


13. A catalyst is a substance that allows a reaction to proceed faster by lowering the energy of activation ($E_{act}$).

15. Since the reaction is endothermic, energy is required for the reaction to proceed in the forward direction. Thus, the $E_{act}$ is greater in the forward direction.

(Note: If the rate of reaction doubles, the amount of time is half as long.)

17. For every 10°C increase in temperature the time is halved.
Since there are three 10°C increases in temperature, the rate is
$\frac{1}{2} \times \frac{1}{2} \times \frac{1}{2} = \frac{1}{8}$ times faster.

$120 \text{ seconds} \times \frac{1}{8} = 15 \text{ seconds}$

Thus, at 50°C the reaction takes 15 seconds.

19. rate = $k$ [A] [B]

(a) rate = $k$ [A] × 2 [B]
If [A] increases twofold, the rate doubles.
Since the rate doubles, the amount of time is halved.
60.0 seconds/2 = 30.0 seconds.

(b) rate = $\frac{k\,[A]\,[B]}{2}$
If [A] decreases twofold, then the rate is halved.
Since the rate is halved, the amount of time is doubled.
60.0 seconds × 2 = 120 seconds

(c) rate = $k$ [A] [B] × 4
If [B] increases fourfold, the rate is quadrupled.
Since the rate is quadrupled, the amount of time decreases fourfold.
60.0 seconds/4 ≐ 15.0 seconds

(d) rate = $\frac{k\,[A]\,[B]}{3}$
If [B] decreases threefold, the rate decreases threefold.
Since the rate decreases threefold, the amount of time is tripled.
60.0 seconds × 3 = 180 seconds

21.* rate = $k\,[NO]^2\,[Cl_2]$

(a) If [NO] increases from 1.00 *M* to 1.50 *M*, then [NO] increases by a factor of (1.50/1.00) and the rate increases by a factor of $(1.50)^2$. Thus, the rate increases 2.25 times.

* continued on next page

21. (b) If [NO] decreases from 0.950 $M$ to 0.750 $M$, then [NO] decreases by a factor of (0.750/0.950) and the rate decreases by a factor of $(0.750/0.950)^2$. Thus, the rate decreases 0.623 times.

(c) If $[Cl_2]$ increases from 1.00 $M$ to 3.00 $M$, then $[Cl_2]$ increases by a factor of (3.00/1.00) and the rate increases by a factor of 3.00. Thus, the rate increases 3.00 times.

(d) If $[Cl_2]$ decreases from 0.650 $M$ to 0.125 $M$, then $[Cl_2]$ decreases by a factor of (0.125/0.650) and the rate decreases by a factor of (0.125/0.650). Thus, the rate decreases 0.192 times.

**Section 16.4** *Law of Chemical Equilibrium*

23. (a) The rate of the forward reaction is measured by the rate of change in the decrease in concentration of reactant(s) in a given unit of time.

$$\text{rate of a forward reaction} = \frac{\text{decrease [reactant]}}{\text{unit time}}$$

(b) The rate of the forward reaction is measured by the rate of change in the increase in concentration of product(s) in a given unit of time.

$$\text{rate of a forward reaction} = \frac{\text{increase [product]}}{\text{unit time}}$$

25. Both (a) and (b) are true statements regarding the general equilibrium expression.

27.

| | Reaction | General Eq. Constant Expression |
|---|---|---|
| (a) | $2\,A \rightleftarrows C$ | $K_{eq} = \frac{[C]}{[A]^2}$ |
| (b) | $A + 2\,B \rightleftarrows 3\,C$ | $K_{eq} = \frac{[C]^3}{[A]\,[B]^2}$ |
| (c) | $2\,A + 3\,B \rightleftarrows 4\,C + D$ | $K_{eq} = \frac{[C]^4\,[D]}{[A]^2\,[B]^3}$ |

**Section 16.5** *Concentration Equilibrium Constant, $K_c$*

29. Since the amount of any solid present has no effect on the equilibrium of a gaseous state reaction, any substance in the solid state will not appear in the equilibrium expression for a gaseous state reaction.

31. Reaction | Eq. Constant Expression

(a) $H_{2(g)} + F_{2(g)} \rightleftarrows 2\,HF_{(g)}$ $\quad K_c = \dfrac{[HF]^2}{[H_2]\,[F_2]}$

(b) $4\,NH_{3(g)} + 7\,O_{2(g)} \rightleftarrows 4\,NO_{2(g)} + 6\,H_2O_{(g)}$ $\quad K_c = \dfrac{[NO_2]^4\,[H_2O]^6}{[NH_3]^4\,[O_2]^7}$

(c) $ZnCO_{3(s)} \rightleftarrows ZnO_{(s)} + CO_{2(g)}$ $\quad K_c = [CO^2]$

33. $2\,SO_{2(g)} + O_{2(g)} \rightleftarrows 2\,SO_{3(g)}$

$[SO_2] = 1.75$
$[O_2] = 1.50$
$[SO_3] = 2.25$

$$K_c = \frac{[SO_3]^2}{[SO_2]^2\,[O_2]} = \frac{(2.25)^2}{(1.75)^2\,(1.50)} = 1.10$$

## Section 16.6 *Gaseous State Equilibria Shifts*

35. (a) On hot days, the heat shifts the equilibrium to the right creating more $NO_2$, and less $N_2O_4$.

(b) On cool, overcast days, the lack of heat shifts the equilibrium to the left creating more $N_2O_4$, and less $NO_2$.

37.

| | Stress | Shift | | Stress | Shift |
|---|---|---|---|---|---|
| (a) | $[CO_2]$ decreases | left | (b) | [CO] decreases | right |
| (c) | solid charcoal is added | none | (d) | $CO_{2(g)}$ is added | right |
| (e) | $N_{2(g)}$ is added | none | (f) | temperature increases | right |
| (g) | pressure increases | left | (h) | volume increases | right |

39.

| | Stress | Shift | | Stress | Shift |
|---|---|---|---|---|---|
| (a) | $[SO_2]$ decreases | left | (b) | $[SO_3]$ decreases | right |
| (c) | $[NO_2]$ increases | right | (d) | [NO] increases | left |
| (e) | temperature decreases | right | (f) | pressure increases | none |
| (g) | volume increases | none | (h) | a catalyst is added | none |

## Section 16.7 *Aqueous Solution Equilibria*

41. Forces That Drive a Reversible Reaction to Completion

(1) formation of a gas
(2) formation of an insoluble precipitate
(3) formation of a weak electrolyte

43.

| | Product | Classification | Effect on Equilibrium |
|---|---|---|---|
| (a) | $Li_2SO_{4(aq)}$ | strong electrolyte | no effect |
| (b) | $BaSO_{4(s)}$ | insoluble precipitate | drives reaction to completion |
| (c) | $H_2SO_{3(aq)}$ | weak electrolyte | drives reaction to completion |
| (d) | $H_2SO_{4(aq)}$ | strong electrolyte | no effect |
| (e) | $KOH_{(aq)}$ | strong electrolyte | no effect |
| (f) | $NH_{3(aq)}$ | weak electrolyte | drives reaction to completion |

45. Product(s) Responsible For Driving Reaction to Completion

(a) $H_2O_{(l)}$ (weak electrolyte) and $CO_{2(g)}$ (gas)
(b) $H_2O_{(l)}$ (weak electrolyte)
(c) $PbCrO_{4(s)}$ (insoluble precipitate)

## Section 16.8 *Ionization Constant, $K_i$*

47.

| | Reaction | Eq. Constant Expression |
|---|---|---|
| (a) | $HCO_2H_{(aq)} \rightleftarrows H^+_{(aq)} + CO_2H^-_{(aq)}$ | $K_i = \dfrac{[H^+]\,[CO_2H^-]}{[HCO_2H]}$ |
| (b) | $H_2C_2O_{4(aq)} \rightleftarrows H^+_{(aq)} + HC_2O_4^-{}_{(aq)}$ | $K_i = \dfrac{[H^+]\,[HC_2O_4^-]}{[H_2C_2O_4]}$ |
| (c) | $H_3C_6H_5O_{7(aq)} \rightleftarrows H^+_{(aq)} + H_2C_6H_5O_7^-{}_{(aq)}$ | $K_i = \dfrac{[H^+]\,[H_2C_6H_5O_7^-]}{[H_3C_6H_5O_7]}$ |

49. $HNO_{2(aq)} \rightleftarrows H^+_{(aq)} + NO_2^-{}_{(aq)}$

$[HNO_2] = 0.125$
$[H+] = 7.5 \times 10^{-3}$
$[NO_2^-] = 7.5 \times 10^{-3}$

$$K_i = \frac{[H^+]\,[NO_2^-]}{[HNO_2]} = \frac{(7.5 \times 10^{-3})\,(7.5 \times 10^{-3})}{(0.125)} = 4.5 \times 10^{-4}$$

51. $HF_{(aq)} \rightleftarrows H^+_{(aq)} + F^-_{(aq)}$

$[HF] = 0.139$
$[H^+] = 10^{-pH} = 10^{-2.00} = 0.010$
$[F^-] = 0.010$

$$K_i = \frac{[H^+]\,[F^-]}{[HF]} = \frac{(0.010)\,(0.010)}{(0.139)} = 7.2 \times 10^{-4}$$

## Section 16.9 *Weak Acid-Base Equilibria Shifts*

53.

| | Stress | Shift | | Stress | Shift |
|---|---|---|---|---|---|
| (a) | increase [HF] | right | (b) | increase $[H^+]$ | left |
| (c) | decrease [HF] | left | (d) | decrease $[F^-]$ | right |
| (e) | add NaF solid | left | (f) | add HCl gas | left |
| (g) | add NaOH solid | right | (h) | increase pH | right |

55.

| | Stress | Shift | | Stress | Shift |
|---|---|---|---|---|---|
| (a) | increase $[HC_2H_3O_2]$ | right | (b) | increase $[H^+]$ | left |
| (c) | decrease $[HC_2H_3O_2]$ | left | (d) | decrease $[C_2H_3O_2^-]$ | right |
| (e) | add $NaC_2H_3O_2$ solid | left | (f) | add NaCl solid | none |
| (g) | add NaOH solid | right | (h) | increase pH | right |

## Section 16.10 *Solubility Product Equilibrium Constant, $K_{sp}$*

57.

| | Reaction | Solubility Product Expression |
|---|---|---|
| (a) | $MnCO_{3(s)} \rightleftarrows Mn^{2+}{}_{(aq)} + CO_3{}^{2-}{}_{(aq)}$ | $K_{sp} = [Mn^{2+}]\,[CO_3{}^{2-}]$ |
| (b) | $Cd(CN)_{2(s)} \rightleftarrows Cd^{2+}{}_{(aq)} + 2\ CN^-{}_{(aq)}$ | $K_{sp} = [Cd^{2+}]\,[CN^-]^2$ |
| (c) | $Sb_2S_{3(s)} \rightleftarrows 2\ Sb^{3+}{}_{(aq)} + 3\ S^{2-}{}_{(aq)}$ | $K_{sp} = [Sb^{3+}]^2\,[S^{2-}]^3$ |

59. $CoS_{(s)} \rightleftarrows Co^{2+}{}_{(aq)} + S^{2-}{}_{(aq)}$

$[Co^{2+}] = 7.7 \times 10^{-11}$
$[S^{2-}] = 7.7 \times 10^{-11}$

$$K_{sp} = [Co^{2+}]\,[S^{2-}] = (7.7 \times 10^{-11})\,(7.7 \times 10^{-11}) = 5.9 \times 10^{-21}$$

61. $Zn_3(PO_4)_{2(s)} \rightleftarrows 3\ Zn^{2+}{}_{(aq)} + 2\ PO_4{}^{3-}{}_{(aq)}$

$[Zn^{2+}] = 1.5 \times 10^{-7}$

$$[PO_4{}^{3-}] = \frac{2}{3}[Zn^{2+}] = \frac{2}{3}(1.5 \times 10^{-7}) = 1.0 \times 10^{-7}$$

$$K_{sp} = [Zn^{2+}]^3\,[PO_4{}^{3-}]^2 = (1.5 \times 10^{-7})^3\,(1.0 \times 10^{-7})^2 = 3.4 \times 10^{-35}$$

63. Let **S** = solubility

$CaCO_{3(s)} \rightleftarrows Ca^{2+}{}_{(aq)} + CO_3{}^{2-}{}_{(aq)}$

**S** **S**

$[Ca^{2+}]\,[CO_3{}^{2-}] = (\mathbf{S})(\mathbf{S}) = K_{sp} = 3.8 \times 10^{-9}$

$\mathbf{S}^2 = 3.8 \times 10^{-9}$

$\mathbf{S} = (3.8 \times 10^{-9})^{1/2} = 6.2 \times 10^{-5}$

Let **S** = solubility

$CaC_2O_{4(s)} \rightleftarrows Ca^{2+}{}_{(aq)} + C_2O_4{}^{2-}{}_{(aq)}$

**S** **S**

$[Ca^{2+}]\,[C_2O_4{}^{2-}] = (\mathbf{S})(\mathbf{S}) = K_{sp} = 2.3 \times 10^{-9}$

$\mathbf{S}^2 = 2.3 \times 10^{-9}$

$\mathbf{S} = (2.3 \times 10^{-9})^{1/2} = 4.8 \times 10^{-5}$

Since the solubility (**S**) of $CaCO_3$ is greater, it is more soluble.

**Section 16.11** *Dissociation Equilibria Shifts*

65.

| | Stress | Shift | | Stress | Shift |
|---|---|---|---|---|---|
| (a) | increase $[Ca^{2+}]$ | left | (b) | increase $[PO_4{}^{3-}]$ | left |
| (c) | decrease $[Ca^{2+}]$ | right | (d) | decrease $[PO_4{}^{3-}]$ | right |
| (e) | add solid $Ca(NO_3)_2$ | left | (f) | add solid $KNO_3$ | none |

67.

| | Stress | Shift | | Stress | Shift |
|---|---|---|---|---|---|
| (a) | increase $[Cu^{2+}]$ | left | (b) | increase $[OH^-]$ | left |
| (c) | decrease $[Cu^{2+}]$ | right | (d) | decrease $[OH^-]$ | right |
| (e) | add solid $Cu(OH)_2$ | none | (f) | add solid NaOH | left |
| (g) | add solid NaCl | none | (h) | add $H^+$ | right |

**General Exercises**

69. (a) The pendulum is in constant motion and reverses its swing in opposing directions.

(b) A solar calculator requires energy to operate (discharging). The reverse process (charging) is ongoing while exposed to light. In addition, the amount of charge the solar calculator takes in is equal to the amount of discharge it can give up.

71. rate = k $[NO_2]^2$

The reaction rate depends only on $[NO_2]$. Later, you may learn that a reaction may require more than one step. The slowest step dictates the rate expression. Apparently, the rate of this reaction does not depend on [CO] because CO molecules are not involved in the slowest step.

73. $N_2O_4(g) \rightleftarrows 2\,NO_2(g)$

$[N_2O_4] = 4.5 \times 10^{-5}$
$[NO_2] = 3.0 \times 10^{-3}$

$$K_c = \frac{[NO_2]^2}{[N_2O_4]} = \frac{(3.0 \times 10^{-3})^2}{(4.5 \times 10^{-5})} = 0.20$$

75. One possible answer is that the stress of final exams causes a shift toward concentration on course work and away from extracurricular activities.

77. (a) $ZnS(s) \rightleftarrows Zn^{2+}(aq) + S^{2-}(aq)$

$$K_{sp} = [Zn^{2+}]\,[S^{2-}] = \left(\frac{mol}{L}\right)\left(\frac{mol}{L}\right) = (mol/L)^2$$

(b) $Zn(OH)_2(s) \rightleftarrows Zn^{2+}(aq) + 2\,OH^{-}(aq)$

$$K_{sp} = [Zn^{2+}]\,[OH^{-}]^2 = \left(\frac{mol}{L}\right)\left(\frac{mol}{L}\right)^2 = (mol/L)^3$$

79. $Cr(OH)_3(s) \rightleftarrows Cr^{3+}(aq) + 3\,OH^{-}(aq)$

$pOH = 14.00 - 7.42 = 6.58$
$[OH^-] = 10^{-pOH} = 10^{-6.58} = 2.6 \times 10^{-7}$

$$[Cr^{3+}] = \frac{1}{3}[OH^-] = \frac{1}{3}(2.6 \times 10^{-7}) = 8.7 \times 10^{-8}$$

$$K_{sp} = [Cr^{3+}]\,[OH^-]^3 = (8.7 \times 10^{-8})\,(2.6 \times 10^{-7})^3 = 1.5 \times 10^{-27}$$

# CHAPTER 17 Oxidation and Reduction

## Section 17.1 *Oxidation Numbers*

1.

| | Metal | Ox No | | Metal | Ox No |
|---|---|---|---|---|---|
| (a) | Mg | 0 | (b) | Mn | 0 |
| (c) | K | 0 | (d) | Zn | 0 |

(Note: All elements in the free state have an oxidation number of zero.)

3.

| | Cation | Ox No | | Cation | Ox No |
|---|---|---|---|---|---|
| (a) | $Sr^{2+}$ | +2 | (b) | $Sc^{3+}$ | +3 |
| (c) | $Ti^{4+}$ | +4 | (d) | $Ag^{+}$ | +1 |

5.

| | Compound | Oxidation Number of Silicon |
|---|---|---|
| (a) | $SiO_2$ | (ox no Si) + 2(ox no O) = 0<br>(ox no Si) + 2(− 2) = 0<br>(ox no Si) + (− 4) = 0<br>ox no Si = + 4 |
| (b) | $Si_2H_6$ | 2(ox no Si) + 6(ox no H) = 0<br>2(ox no Si) + 6(− 1) = 0<br>2(ox no Si) + (− 6) = 0<br>2(ox no Si) = + 6<br>ox no Si = + 3 |
| (c) | $Si_3N_4$ | 3(ox no Si) + 4(ox no N) = 0<br>3(ox no Si) + 4(− 3) = 0<br>3(ox no Si) + (− 12) = 0<br>3(ox no Si) = + 12<br>ox no Si = + 4 |
| (d) | $CaSiO_3$ | (ox no Ca) + (ox no Si) + 3(ox no O) = 0<br>(+ 2) + (ox no Si) + 3(− 2) = 0<br>(ox no Si) + (− 4) = 0<br>ox no Si = + 4 |

7. Ion — Oxidation Number of Carbon

(a) $CO_3^{2-}$

(ox no C) + 3(ox no O) = – 2
(ox no C) + 3(– 2) = – 2
(ox no C) + (– 6) = – 2
ox no C = + 4

(b) $HCO_3^-$

(ox no H) + (ox no C) + 3(ox no O) = – 1
(+ 1) + (ox no C) + 3(– 2) = – 1
(+ 1) + (ox no C) +(– 6) = – 1
ox no C = + 4

(c) $CN^-$

(ox no C) + (ox no N) = – 1
(ox no C) + (– 3) = – 1
ox no C = + 2

(d) $CNO^-$

(ox no C) + (ox no N) + (ox no O) = – 1
(ox no C) + (– 3) + (– 2) = – 1
ox no C = + 4

## Section 17.2 *Oxidation-Reduction Reactions*

9.

| | Statement | Term |
|---|---|---|
| (a) | a redox process characterized by electron loss | oxidation |
| (b) | a redox process characterized by electron gain | reduction |

11.* (a) Mn: 0 → +4; O: 0 → –2

$Mn_{(s)} + O_{2(g)} \rightarrow MnO_{2(s)}$

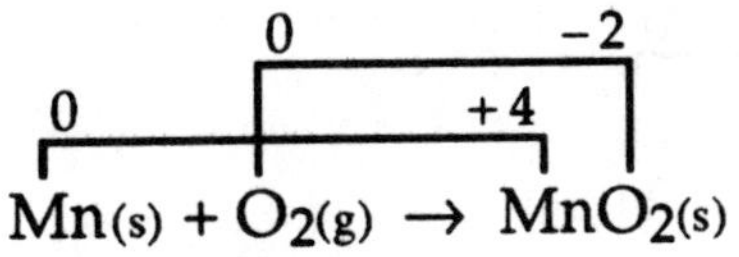

Oxidized: Mn — Reduced: $O_2$

(b) S: 0 → +4; O: 0 → –2

$S_{(s)} + O_{2(g)} \rightarrow SO_{2(g)}$

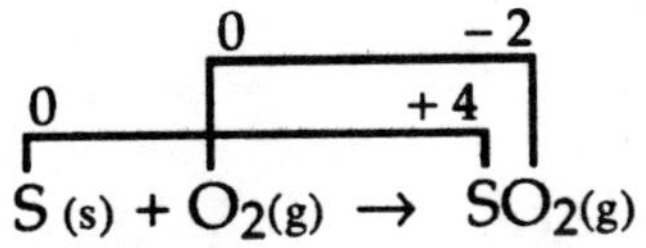

Oxidized: S — Reduced: $O_2$

(c) Cd: 0 → +2; F: 0 → –1

$Cd_{(s)} + F_{2(g)} \rightarrow CdF_{2(s)}$

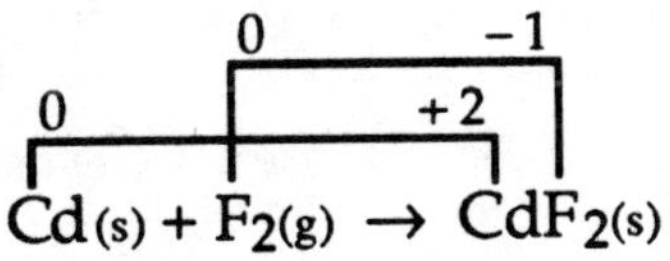

Oxidized: Cd — Reduced: $F_2$

---

* continued on next page

11. (d) $Sr_{(s)} + 2\,H_2O_{(l)} \rightarrow Sr(OH)_{2(aq)} + H_{2(g)}$

(Oxidation numbers: Sr 0 → +2; H +1 → 0)

Oxidized: Sr Reduced: $H_2O$

(e) $Pb_{(s)} + CuSO_{4(aq)} \rightarrow PbSO_{4(aq)} + Cu_{(s)}$

(Oxidation numbers: Pb 0 → +2; Cu +2 → 0)

Oxidized: Pb Reduced: $CuSO_4$

13. (a) $CuO_{(s)} + H_{2(g)} \rightarrow Cu_{(s)} + H_2O_{(l)}$

(Oxidation numbers: Cu +2 → 0; H 0 → +1)

Oxidized: $H_2$ Reduced: CuO

(b) $Cl_{2(g)} + 2\,KBr_{(aq)} \rightarrow Br_{2(l)} + 2\,KCl_{(aq)}$

(Oxidation numbers: Cl 0 → −1; Br −1 → 0)

Oxidized: KBr Reduced: $Cl_2$

(c) $Ca_{(s)} + 2\,H_2O_{(l)} \rightarrow Ca(OH)_{2(aq)} + H_{2(g)}$

(Oxidation numbers: Ca 0 → +2; H +1 → 0)

Oxidized: Ca Reduced: $H_2O$

(d) $Mg_{(s)} + 2\,HCl_{(aq)} \rightarrow MgCl_{2(aq)} + H_{2(g)}$

(Oxidation numbers: Mg 0 → +2; H +1 → 0)

Oxidized: Mg Reduced: HCl

(e) $PbO_{(s)} + CO_{(g)} \rightarrow Pb_{(s)} + CO_{2(g)}$

(Oxidation numbers: Pb +2 → 0; C +2 → +4)

Oxidized: CO Reduced: PbO

15. (a)

+3 0

0 +3

$Al_{(s)} + Cr^{3+}_{(aq)} \rightarrow Al^{3+}_{(aq)} + Cr_{(s)}$

Oxidized: Al Reduced: $Cr^{3+}$

(b)

−1 0

0 −1

$F_{2(g)} + 2\ Cl^{-}_{(aq)} \rightarrow 2\ F^{-}_{(aq)} + Cl_{2(g)}$

Oxidized: $Cl^{-}$ Reduced: $F_2$

(c)

+4 +6

+3 +2

$H_2O_{(l)} + 2\ Fe^{3+}_{(aq)} + SO_3{}^{2-}_{(aq)} \rightarrow 2\ Fe^{2+}_{(aq)} + SO_4{}^{2-}_{(aq)} + 2\ H^{+}_{(aq)}$

Oxidized: $SO_3{}^{2-}$ Reduced: $Fe^{3+}$

(d)

+2 +1

+2 +4

$Sn^{2+}_{(aq)} + 2\ Hg^{2+}_{(aq)} \rightarrow Sn^{4+}_{(aq)} + Hg_2{}^{2+}_{(aq)}$

Oxidized: $Sn^{2+}$ Reduced: $Hg^{2+}$

(e)

+1 0

+2 +3

$Cr^{2+}_{(aq)} + AgI_{(s)} \rightarrow Cr^{3+}_{(aq)} + Ag_{(s)} + I^{-}_{(aq)}$

Oxidized: $Cr^{2+}$ Reduced: AgI

17.* (a)

+2 0

−1 0

$2\ Br^{-}_{(aq)} + Pt^{2+}_{(aq)} \rightarrow Br_{2(l)} + Pt_{(s)}$

Oxidized: $Br^{-}$ Reduced: $Pt^{2+}$

(b)

0 +4

+4 0

$TeO_{2(s)} + 4\ H^{+}_{(aq)} + Zr_{(s)} \rightarrow Te_{(s)} + Zr^{4+}_{(aq)} + 2\ H_2O_{(l)}$

Oxidized: Zr Reduced: $TeO_2$

* continued on next page

17. (c) $MnO_{2(s)} + SO_{2(g)} \rightarrow Mn^{2+}_{(aq)} + SO_4^{2-}{}_{(aq)}$

Oxidation numbers: Mn +4 → +2; S +4 → +6

Oxidized: $SO_2$ Reduced: $MnO_2$

(d) $Pb^{4+}_{(aq)} + H_2O_{2(aq)} \rightarrow Pb^{2+}_{(aq)} + O_{2(g)} + 2\,H^+_{(aq)}$

Oxidation numbers: Pb +4 → +2; O −1 → 0

Oxidized: $H_2O_2$ Reduced: $Pb^{4+}$

(e) $H_2O_{(l)} + SO_3^{2-}{}_{(aq)} + I_{2(s)} \rightarrow SO_4^{2-}{}_{(aq)} + 2\,I^-_{(aq)} + 2\,H^+_{(aq)}$

Oxidation numbers: S +4 → +6; I 0 → −1

Oxidized: $SO_3^{2-}$ Reduced: $I_2$

**Section 17.3** *Balancing Redox Equations: Oxidation Number Method*

19. Yes, the total electron loss by oxidation must equal the total electron gain by the reduction process.

21. (a)

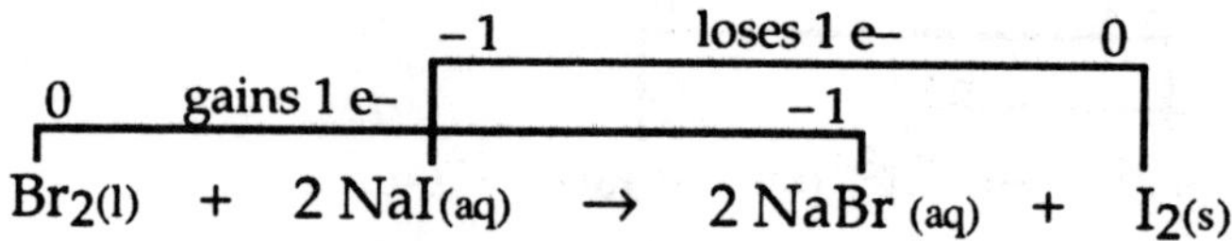


(b)

$2\,PbS_{(s)} + 3\,O_{2(g)} \rightarrow 2\,PbO_{(s)} + 2\,SO_{2(g)}$

O: 0 → −2, gains 2 e–; S: −2 → +4, loses 6 e–

Note: Since PbO and $SO_2$ both contain oxygen with an oxidation number of −2, choose either of the products to follow the gain of electrons from $O_2$.

(c)

$2\,B_2O_{3(s)} + 6\,Cl_{2(g)} \rightarrow 3\,O_{2(g)} + 4\,BCl_{3(aq)}$

Cl: 0 → −1, gains 1 e–; O: −2 → 0, loses 2 e–

23. (a)

-1 loses 1 e– 0

+7 gains 5 e– +2

$$2\,MnO_4^-{}_{(aq)} + 10\,I^-{}_{(aq)} + 16\,H^+{}_{(aq)} \rightarrow 2\,Mn^{2+}{}_{(aq)} + 5\,I_{2(s)} + 8\,H_2O_{(l)}$$

(b)

+6 gains 2 e– +4

0 loses 2 e– +2

$$Cu_{(s)} + 4\,H^+{}_{(aq)} + SO_4^{2-}{}_{(aq)} \rightarrow Cu^{2+}{}_{(aq)} + SO_{2(g)} + 2\,H_2O_{(l)}$$

(c)

-1 gains 1 e– -2

+2 loses 1 e– +3

$$2\,Fe^{2+}{}_{(aq)} + H_2O_{2(aq)} + 2\,H^+{}_{(aq)} \rightarrow 2\,Fe^{3+}{}_{(aq)} + 2\,H_2O_{(l)}$$

## Section 17.4 *Balancing Redox Equations: Ion-Electron Method*

25. (a) $F_{2(g)} + 2\,e^- \rightarrow 2\,F^-{}_{(aq)}$
    (b) $Hg_{(l)} + 2\,Cl^-{}_{(aq)} \rightarrow HgCl_{2(s)} + 2\,e^-$
    (c) $2\,BrO_3^-{}_{(aq)} + 12\,H^+{}_{(aq)} + 10\,e^- \rightarrow Br_{2(l)} + 6\,H_2O_{(l)}$
    (d) $H_2O_{2(aq)} + 2\,H^+{}_{(aq)} + 2\,e^- \rightarrow 2\,H_2O_{(l)}$
    (e) $O_{2(g)} + 2\,H^+{}_{(aq)} + 2\,e^- \rightarrow H_2O_{2(aq)}$

27.* (a) Oxidation: $3\,(Zn \rightarrow Zn^{2+} + 2\,e^-)$
Reduction: $2\,(NO_3^- + 4\,H^+ + 3\,e^- \rightarrow NO + 2\,H_2O)$

$$3\,Zn_{(s)} + 2\,NO_3^-{}_{(aq)} + 8\,H^+{}_{(aq)} \rightarrow 3\,Zn^{2+}{}_{(aq)} + 2\,NO_{(g)} + 4\,H_2O_{(l)}$$

(b) Oxidation: $2\,(4\,H_2O + Mn^{2+} \rightarrow MnO_4^- + 8\,H^+ + 5\,e^-)$
Reduction: $5\,(BiO_3^- + 6\,H^+ + 2\,e^- \rightarrow Bi^{3+} + 3\,H_2O)$

$$2\,Mn^{2+}{}_{(aq)} + 5\,BiO_3^-{}_{(aq)} + 14\,H^+{}_{(aq)} \rightarrow 2\,MnO_4^-{}_{(aq)} + 5\,Bi^{3+}{}_{(aq)} + 7\,H_2O_{(l)}$$

(c) Oxidation: $5\,(SO_3^{2-} + H_2O \rightarrow SO_4^{2-} + 2\,H^+ + 2\,e^-)$
Reduction: $2(MnO_4^- + 8\,H^+ + 5\,e^- \rightarrow Mn^{2+} + 4\,H_2O)$

$$2\,MnO_4^-{}_{(aq)} + 5\,SO_3^{2-}{}_{(aq)} + 6\,H^+{}_{(aq)} \rightarrow 2\,Mn^{2+}{}_{(aq)} + 5\,SO_4^{2-}{}_{(aq)} + 3\,H_2O_{(l)}$$

---

* continued on next page

27. (d) Oxidation: $5\,(Sn^{2+} \rightarrow Sn^{4+} + 2\,e^-)$

Reduction: $2\,IO_3^- + 12\,H^+ + 10\,e^- \rightarrow I_2 + 6\,H_2O$

$$5\,Sn^{2+}(aq) + 2\,IO_3^-(aq) + 12\,H^+(aq) \rightarrow 5\,Sn^{4+}(aq) + I_2(s) + 6\,H_2O(l)$$

(e) Oxidation: $AsO_3^{3-} + H_2O \rightarrow AsO_4^{3-} + 2\,H^+ + 2\,e^-$

Reduction: $Br_2 + 2\,e^- \rightarrow 2\,Br^-$

$$AsO_3^{3-}(aq) + Br_2(l) + H_2O(l) \rightarrow AsO_4^{3-}(aq) + 2\,Br^-(aq) + 2\,H^+(aq)$$

29. (a) Oxidation: $Cl_2 + 2\,H_2O \rightarrow 2\,HOCl + 2\,H^+ + 2\,e^-$

Reduction: $Cl_2 + 2\,e^- \rightarrow 2\,Cl^-$

$$Cl_2(g) + H_2O(l) \rightarrow Cl^-(aq) + HOCl(aq) + H^+(aq)$$

(b) Oxidation: $HNO_2 + H_2O \rightarrow NO_3^- + 3\,H^+ + 2\,e^-$

Reduction: $2\,(HNO_2 + H^+ + 1\,e^- \rightarrow NO + H_2O)$

$$3\,HNO_2(aq) \rightarrow NO_3^-(aq) + 2\,NO(g) + H_2O(l) + H^+(aq)$$

(c) Oxidation: $2\,(MnO_4^{2-} \rightarrow MnO_4^- + 1\,e^-)$

Reduction: $MnO_4^{2-} + 4\,H^+ + 2\,e^- \rightarrow MnO_2 + 2\,H_2O$

$$3\,MnO_4^{2-}(aq) + 4\,H^+(aq) \rightarrow 2\,MnO_4^-(aq) + MnO_2(s) + 2\,H_2O(l)$$

## Section 17.5 *Predicting Spontaneous Redox Reactions*

31.

| | Substances | Most Likely to be Reduced |
|---|---|---|
| (a) | $Au^{3+}(aq)$ or $Hg^{2+}(aq)$ | $Au^{3+}(aq)$ |
| (b) | $H^+(aq)$ or $H_2(g)$ | $H^+(aq)$ |
| (c) | $Fe^{2+}(aq)$ or $I_2(s)$ | $I_2(s)$ |
| (d) | $Al^{3+}(aq)$ or $Br_2(l)$ | $Br_2(l)$ |

33.

| | Substances | Stronger Oxidizing Agent |
|---|---|---|
| (a) | $F_2(g)$ or $Cl_2(g)$ | $F_2(g)$ |
| (b) | $Ag^+(aq)$ or $Br_2(l)$ | $Br_2(l)$ |
| (c) | $Cu^{2+}(aq)$ or $H^+(aq)$ | $Cu^{2+}(aq)$ |
| (d) | $Mg^{2+}(aq)$ or $Mn^{2+}(aq)$ | $Mn^{2+}(aq)$ |

35. (a) spontaneous
(b) nonspontaneous
(c) spontaneous
(d) nonspontaneous
(e) spontaneous

37. **Balanced Redox Reactions**

(a) $Mg_{(s)} + Sn^{2+}{}_{(aq)} \rightarrow Mg^{2+}{}_{(aq)} + Sn_{(s)}$

(b) $2\ H^{+}{}_{(aq)} + 2\ I^{-}{}_{(aq)} \rightarrow H_{2(g)} + I_{2(s)}$

(c) $2\ H^{+}{}_{(aq)} + Ni_{(s)} \rightarrow H_{2(g)} + Ni^{2+}{}_{(aq)}$

(d) $Hg^{2+}{}_{(aq)} + 2\ Br^{-}{}_{(aq)} \rightarrow Hg_{(l)} + Br_{2(l)}$

(e) $2\ Fe^{3+}{}_{(aq)} + 2\ I^{-}{}_{(aq)} \rightarrow 2\ Fe^{2+}{}_{(aq)} + I_{2(s)}$

## Section 17.6 *Voltaic Cells*

39.

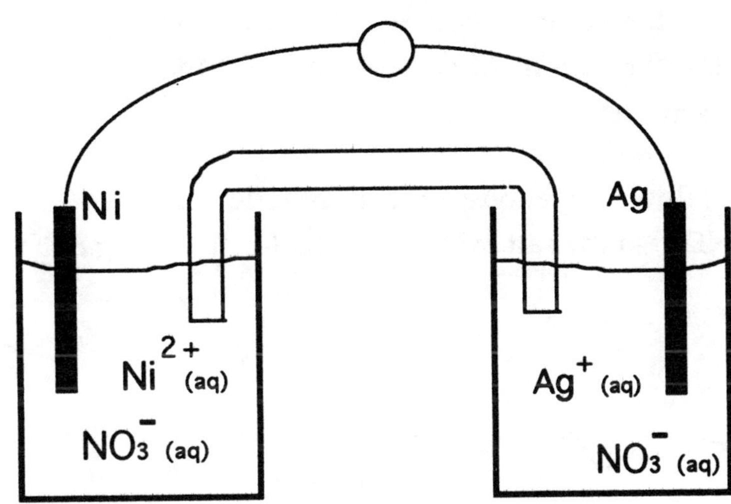


$Ni_{(s)} + 2\ AgNO_{3(aq)} \rightarrow 2\ Ag_{(s)} + Ni(NO_3)_{2(aq)}$

41.

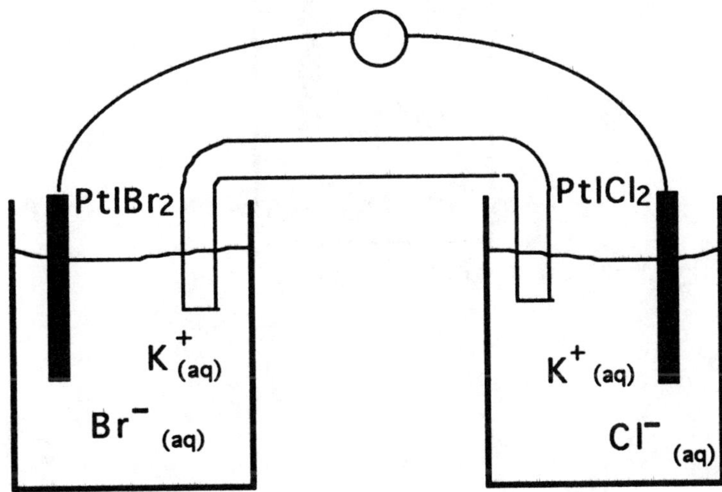


$Cl_{2(g)} + 2\ KBr_{(aq)} \rightarrow 2\ KCl_{(aq)} + Br_{2(l)}$

43. $Sn_{(s)} + Cu^{2+}_{(aq)} \rightarrow Cu_{(s)} + Sn^{2+}_{(aq)}$

(a) oxidation half-cell reaction: $Sn_{(s)} \rightarrow Sn^{2+}_{(aq)} + 2\ e^-$

(b) reduction half-cell reaction: $Cu^{2+}_{(aq)} + 2\ e^- \rightarrow Cu_{(s)}$

(c) anode: Sn electrode
cathode: Cu electrode

(d) direction of $e^-$ flow: Sn anode → Cu cathode

(e) direction of $SO_4^{2-}$ in the salt bridge: Cu half-cell → Sn half-cell

45. $Mg_{(s)} + Mn(NO_3)_{2(aq)} \rightarrow Mn_{(s)} + Mg(NO_3)_{2(aq)}$

(a) oxidation half-cell reaction: $Mg_{(s)} \rightarrow Mg^{2+}_{(aq)} + 2\ e^-$

(b) reduction half-cell reaction: $Mn^{2+}_{(aq)} + 2\ e^- \rightarrow Mn_{(s)}$

(c) anode: Mg electrode
cathode: Mn electrode

(d) direction of $e^-$ flow: Mg anode → Mn cathode

(e) direction of $NO_3^-$ in the salt bridge: Mn half-cell → Mg half-cell

**Section 17.7** *Electrolytic Cells*

47.

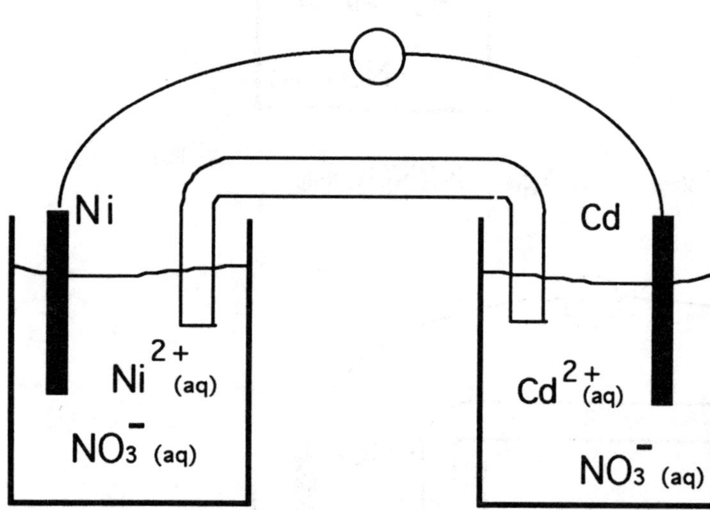


$Ni_{(s)} + Cd(NO_3)_{2(aq)} \rightarrow Cd_{(s)} + Ni(NO_3)_{2(aq)}$

49.

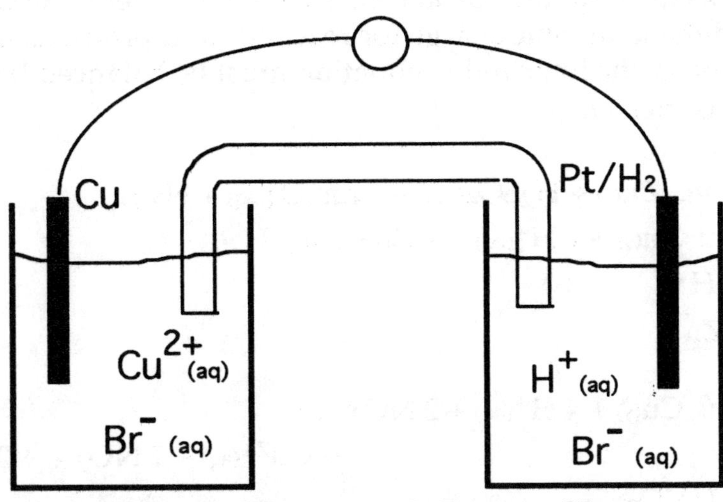


$Cu_{(s)} + 2\ HBr_{(aq)} \rightarrow CuBr_{2(aq)} + H_{2(g)}$

51. $Ni_{(s)} + Fe^{2+}_{(aq)} \rightarrow Fe_{(s)} + Ni^{2+}_{(aq)}$

(a) oxidation half-cell reaction: $Ni_{(s)} \rightarrow Ni^{2+}_{(aq)} + 2\ e^-$

(b) reduction half-cell reaction: $Fe^{2+}_{(aq)} + 2\ e^- \rightarrow Fe_{(s)}$

(c) anode: Ni electrode
cathode: Fe electrode

(d) direction of $e^-$ flow: Ni anode $\rightarrow$ Fe cathode

(e) direction of $SO_4^{2-}$ in the salt bridge: Fe half-cell $\rightarrow$ Ni half-cell

53. $Cr_{(s)} + Al(C_2H_3O_2)_{3(aq)} \rightarrow Al_{(s)} + Cr(C_2H_3O_2)_{3(aq)}$

(a) oxidation half-cell reaction: $Cr_{(s)} \rightarrow Cr^{3+}_{(aq)} + 3\ e^-$

(b) reduction half-cell reaction: $Al^{3+}_{(aq)} + 3\ e^- \rightarrow Al_{(s)}$

(c) anode: Cr electrode
cathode: Al electrode

(d) direction of $e^-$ flow: Cr anode $\rightarrow$ Al cathode

(e) direction of $C_2H_3O_2^-$ is the salt bridge: Al half-cell $\rightarrow$ Cr half-cell

## General Exercises

55.

| Compound | Oxidation Number of Sulfur |
|---|---|
| $Na_2S_2O_3$ | 2(ox no Na) + 2(ox no S) + 3(ox no O) = 0 |
| | 2(+ 1) + 2(ox no S) + 3(− 2) = 0 |
| | 2(ox no S) + (− 4) = 0 |
| | 2(ox no S) = + 4 |
| | ox no S = + 2 |

57. For an ionic redox equation to be balanced, the total number of atoms of each element and the total ionic charge for reactants and products must be equal. In other words, the ionic redox equation must be balanced both atomically and electrically.

59. molecular equation: $Zn_{(s)} + H_2SO_{4(aq)} \rightarrow ZnSO_{4(aq)} + H_{2(g)}$
net ionic equation: $Zn_{(s)} + 2\ H^+_{(aq)} \rightarrow Zn^{2+}_{(aq)} + H_{2(g)}$
oxidizing agent: $H^+$
reducing agent: Zn

61. net ionic equation: $Cu_{(s)} + 4\ H^+_{(aq)} + 2\ NO_3^-{}_{(aq)}$
$\rightarrow Cu^{2+}_{(aq)} + 2\ NO_{2(g)} + 2\ H_2O_{(l)}$
oxidizing agent: $HNO_3$
reducing agent: Cu

63. Advantages
(1) Fuel cells are more energy efficient.
(2) Fuels cells do not produce atmospheric pollution.

# Advanced Chemical Calculations

CHAPTER 18

**Section 18.1** *Advanced Problem Solving*

1. (a) The first step in solving any problem is to determine the unknown quantity (for example, mass in grams, volume in liters, etc...).

   (b) The second step in solving any problem is to determine what given quantity is related to the unknown quantity that is to be calculated.

   (c) The two quantities that are connected by a strategy map are the given quantity and the unknown quantity; for example,

$$\text{L } N_2 \text{ (STP)} \rightarrow \text{mol } SO_2 \rightarrow \text{g } SO_2$$

   (d) A concept map is more general than a strategy map and simply relates two quantities; for example,

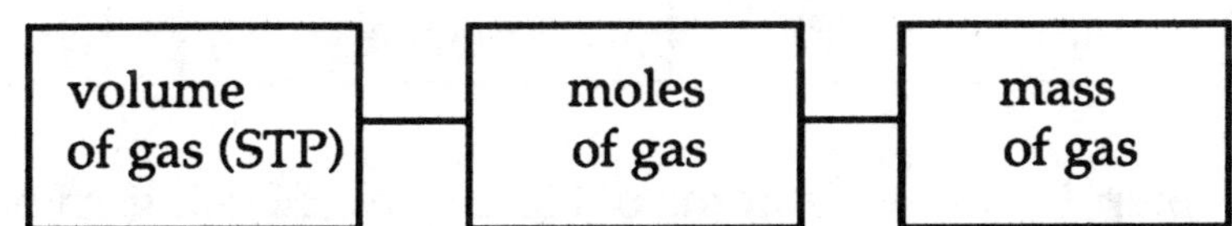


   (e) The process of writing systematic steps for solving a problem is known as an algorithm. An algorithm writes out a series of sequential steps in a calculation operation.

   (f) The process of forming mental pictures to make a problem more concrete is known as visualization.

3.*

| | Unit | Symbol | Physical Quantity |
|---|---|---|---|
| (a) | gram | g | mass |
| (b) | second | s | time |
| (c) | cubic centimeter | $cm^3$ | volume |
| (d) | inch | in. | length |

* continued on next page

3.

| | Unit | Symbol | Physical Quantity |
|---|---|---|---|
| (e) | pound | lb | weight |
| (f) | Kelvin | K | temperature |
| (g) | kilogram | kg | mass |
| (h) | milliliter | mL | volume |
| (i) | moles per liter | *M* | concentration |
| (j) | degree Celsius | °C | temperature |
| (k) | decimeter | dm | length |
| (l) | grams per liter | g/L | density |
| (m) | quart | qt | volume |
| (n) | microsecond | μs | time |

5. A possible concept map for the integrated software program is:

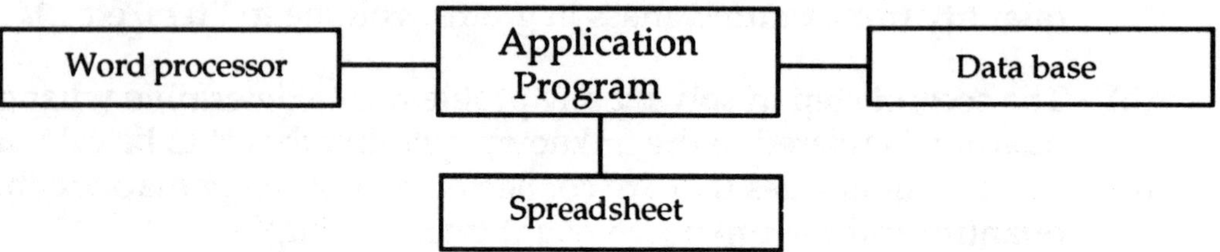


7. (a)

(b)

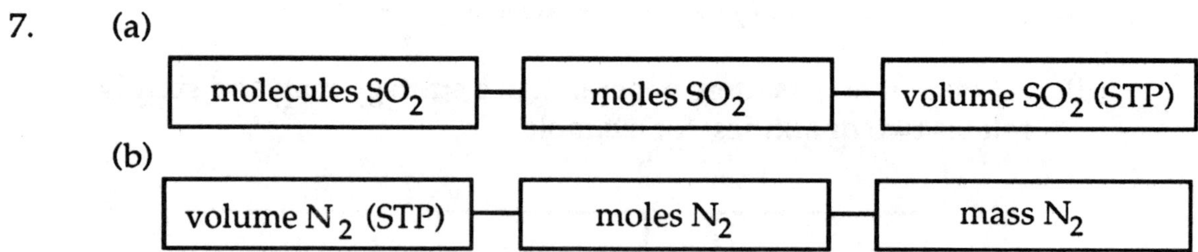


9. (a) Step 1: Convert atoms of He to mol He using Avogadro's number.
Step 2: Convert mol He to volume of He (STP) using molar volume.

(b) Step 1: Convert the volume of $Cl_2$ (STP) to moles of $Cl_2$ using the molar volume.
Step 2: Convert the moles of $Cl_2$ to mass of $Cl_2$ using the molar mass.

11. (a)

(b)

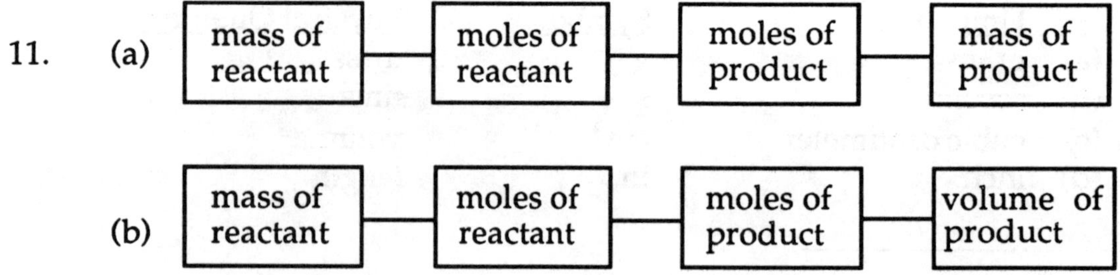

13. Consider the general equation: a A + b B → c C + d D

(a) Step 1: Convert the mass of reactant A to the moles of A using the molar mass of A.
Step 2: Convert the moles of reactant A to the moles of product C using the coefficients from the balanced chemical equation.
Step 3: Convert the moles of product C to the mass of C using the molar mass of C.

(b) Step 1: Convert the mass of reactant A to the moles of A using the molar mass of A.
Step 2: Convert the moles of reactant A to the moles of gaseous product D using the coefficients from the balanced equation.
Step 3: Convert the moles of product D to the volume of D (at STP) using the molar volume concept.

15. (a) $$1.550\ \cancel{g\ Cr_2O_3} \times \frac{1\ \cancel{mol\ Cr_2O_3}}{152.0\ \cancel{g\ Cr_2O_3}} \times \frac{2\ \cancel{mol\ Cr}}{1\ \cancel{mol\ Cr_2O_3}} \times \frac{52.0\ g\ Cr}{1\ \cancel{mol\ Cr}} = g\ Cr$$

$$\frac{3\ \cancel{g\ Cr_2O_3}}{2} \times \frac{1\ \cancel{mol\ Cr_2O_3}}{150\ \cancel{g\ Cr_2O_3}} \times \frac{2\ \cancel{mol\ Cr}}{1\ \cancel{mol\ Cr_2O_3}} \times \frac{50\ g\ Cr}{1\ \cancel{mol\ Cr}} = 1\ g\ Cr$$

*Estimated answer:* ~ 1 g Cr

(b) $$42.0\ \cancel{mL\ HNO_3} \times \frac{0.195\ \cancel{mol\ HNO_3}}{1000\ \cancel{mL\ HNO_3}} \times \frac{1\ \cancel{mol\ Ba(OH)_2}}{2\ \cancel{mol\ HNO_3}} \times \frac{1000\ mL\ Ba(OH)_2}{0.105\ \cancel{mol\ Ba(OH)_2}} = mL\ Ba(OH)_2$$

$$40\ \cancel{mL\ HNO_3} \times \frac{0.2\ \cancel{mol\ HNO_3}}{1000\ \cancel{mL\ HNO_3}} \times \frac{1\ \cancel{mol\ Ba(OH)_2}}{2\ \cancel{mol\ HNO_3}} \times \frac{1000\ mL\ Ba(OH)_2}{0.1\ \cancel{mol\ Ba(OH)_2}} = 40\ mL\ Ba(OH)_2$$

*Estimated answer:* ~ 40 mL $Ba(OH)_2$

(c) $$21.5\ \cancel{mL\ BiCl_3} \times \frac{0.115\ \cancel{mol\ BiCl_3}}{1000\ \cancel{mL\ BiCl_3}} \times \frac{3\ \cancel{mol\ Hg_2Cl_2}}{2\ \cancel{mol\ BiCl_3}} \times \frac{472.2\ g\ Hg_2Cl_2}{1\ \cancel{mol\ Hg_2Cl_2}} = g\ Hg_2Cl_2$$

$$20\ \cancel{mL\ BiCl_3} \times \frac{0.1\ \cancel{mol\ BiCl_3}}{1000\ \cancel{mL\ BiCl_3}} \times \frac{3\ \cancel{mol\ Hg_2Cl_2}}{2\ \cancel{mol\ BiCl_3}} \times \frac{500\ g\ Hg_2Cl_2}{1\ \cancel{mol\ Hg_2Cl_2}} = 1.5\ g\ Hg_2Cl_2$$

*Estimated answer:* ~ 1.5 g $Hg_2Cl_2$

**Section 18.2** *Chemical Formula Calculations*

17. (a) $0.0142\ \cancel{\text{mol HCl}} \times \dfrac{36.5\ \text{g HCl}}{1\ \cancel{\text{mol HCl}}} = 0.518\ \text{g HCl}$

(b) $0.0142\ \cancel{\text{mol HCl}} \times \dfrac{22.4\ \text{L HCl}}{1\ \cancel{\text{mol HCl}}} = 0.318\ \text{L HCl}$

(c) $0.0142\ \cancel{\text{mol HCl}} \times \dfrac{6.02 \times 10^{23}\ \text{molecules HCl}}{1\ \cancel{\text{mol HCl}}}$

$= 8.55 \times 10^{21}\ \text{molecules HCl}$

(d) $\dfrac{0.0142\ \text{mol HCl}}{47.5\ \cancel{\text{mL solution}}} \times \dfrac{1000\ \cancel{\text{mL solution}}}{1\ \text{L solution}} = 0.299\ M\ \text{HCl}$

19. (a) $1.00\ \cancel{\text{g HF}} \times \dfrac{1\ \cancel{\text{mol HF}}}{20.0\ \cancel{\text{g HF}}} \times \dfrac{22.4\ \text{L HF}}{1\ \cancel{\text{mol HF}}} = 1.12\ \text{L HF}$

(b) $1.00\ \cancel{\text{g HF}} \times \dfrac{1\ \cancel{\text{mol HF}}}{20.0\ \cancel{\text{g HF}}} \times \dfrac{6.02 \times 10^{23}\ \text{molecules HF}}{1\ \cancel{\text{mol HF}}}$

$= 3.01 \times 10^{22}\ \text{molecules HF}$

(c) $\dfrac{1.00\ \cancel{\text{g HF}}}{100.0\ \cancel{\text{mL solution}}} \times \dfrac{1\ \text{mol HF}}{20.0\ \cancel{\text{g HF}}} \times \dfrac{1000\ \cancel{\text{mL solution}}}{1\ \text{L solution}} = 0.500\ M\ \text{HF}$

21. (a) $1.00\ \cancel{\text{L CO}_2} \times \dfrac{1\ \cancel{\text{mol CO}_2}}{22.4\ \cancel{\text{L CO}_2}} \times \dfrac{44.0\ \text{g CO}_2}{1\ \cancel{\text{mol CO}_2}} = 1.96\ \text{g CO}_2$

(b) $1.00\ \cancel{\text{L CO}_2} \times \dfrac{1\ \cancel{\text{mol CO}_2}}{22.4\ \cancel{\text{L CO}_2}} \times \dfrac{6.02 \times 10^{23}\ \text{molecules CO}_2}{1\ \cancel{\text{mol CO}_2}}$

$= 2.69 \times 10^{22}\ \text{molecules CO}_2$

(c) $\dfrac{1.00\ \cancel{\text{L CO}_2}}{500.0\ \cancel{\text{mL solution}}} \times \dfrac{1\ \text{mol CO}_2}{22.4\ \cancel{\text{L CO}_2}} \times \dfrac{1000\ \cancel{\text{mL solution}}}{1\ \text{L solution}}$

$= 0.0893\ M\ \text{H}_2\text{CO}_3$

23.* (a) $2.22 \times 10^{22}\ \cancel{\text{molecules H}_2\text{S}} \times \dfrac{1\ \cancel{\text{mol H}_2\text{S}}}{6.02 \times 10^{23}\ \cancel{\text{molecules H}_2\text{S}}} \times$

$\dfrac{22.4\ \text{L H}_2\text{S}}{1\ \cancel{\text{mol H}_2\text{S}}} = 0.826\ \text{L H}_2\text{S}$

* continued on next page

23. (b) $2.22 \times 10^{22} \text{ molecules } H_2S \times \dfrac{1 \text{ mol } H_2S}{6.02 \times 10^{23} \text{ molecules } H_2S} \times \dfrac{34.1 \text{ g } H_2S}{1 \text{ mol } H_2S} = 1.26 \text{ g } H_2S$

(c) $\dfrac{2.22 \times 10^{22} \text{ molecules } H_2S}{450.0 \text{ mL solution}} \times \dfrac{1 \text{ mol } H_2S}{6.02 \times 10^{23} \text{ molecules } H_2S} \times \dfrac{1000 \text{ mL solution}}{1 \text{ L solution}} = 0.0819\ M\ H_2S$

**Section 18.3** *Chemical Equation Calculations*

25. $CH_{4(g)} + 2\ O_{2(g)} \rightarrow CO_{2(g)} + 2\ H_2O_{(l)}$

(a) $5.00 \text{ g } H_2O \times \dfrac{1 \text{ mol } H_2O}{18.0 \text{ g } H_2O} \times \dfrac{1 \text{ mol } CH_4}{2 \text{ mol } H_2O} \times \dfrac{16.0 \text{ g } CH_4}{1 \text{ mol } CH_4} = 2.22 \text{ g } CH_4$

(b) $5.00 \text{ g } H_2O \times \dfrac{1 \text{ mol } H_2O}{18.0 \text{ g } H_2O} \times \dfrac{2 \text{ mol } O_2}{2 \text{ mol } H_2O} \times \dfrac{22.4 \text{ L } O_2}{1 \text{ mol } O_2} = 6.22 \text{ L } O_2$

27. $2\ Al_{(s)} + 6\ HCl_{(aq)} \rightarrow 2\ AlCl_{3(aq)} + 3\ H_{2(g)}$

(a) $0.466 \text{ g Al} \times \dfrac{1 \text{ mol Al}}{27.0 \text{ g Al}} \times \dfrac{3 \text{ mol } H_2}{2 \text{ mol Al}} \times \dfrac{22.4 \text{ L } H_2}{1 \text{ mol } H_2} = 0.580 \text{ L } H_2$

(b) $0.466 \text{ g Al} \times \dfrac{1 \text{ mol Al}}{27.0 \text{ g Al}} \times \dfrac{6 \text{ mol HCl}}{2 \text{ mol Al}} \times \dfrac{1 \text{ L solution}}{0.100 \text{ mol HCl}} \times \dfrac{1000 \text{ mL solution}}{1 \text{ L solution}} = 518 \text{ mL HCl solution}$

29. $MgBr_{2(aq)} + 2\ AgNO_{3(aq)} \rightarrow 2\ AgBr_{(s)} + Mg(NO_3)_{2(aq)}$

(a) $50.0 \text{ mL } MgBr_2 \times \dfrac{0.100 \text{ mol } MgBr_2}{1000 \text{ mL } MgBr_2} \times \dfrac{2 \text{ mol } AgNO_3}{1 \text{ mol } MgBr_2} = 0.0100 \text{ mol } AgNO_3$

$\dfrac{0.0100 \text{ mol } AgNO_3}{13.9 \text{ mL solution}} \times \dfrac{1000 \text{ mL solution}}{1 \text{ L solution}} = 0.719\ M\ AgNO_3$

(b) $50.0 \text{ mL } MgBr_2 \times \dfrac{0.100 \text{ mol } MgBr_2}{1000 \text{ mL } MgBr_2} \times \dfrac{2 \text{ mol AgBr}}{1 \text{ mol } MgBr_2} \times \dfrac{187.8 \text{ g AgBr}}{1 \text{ mol AgBr}} = 1.88 \text{ g AgBr}$

**Section 18.4** *Limiting Reactant Concept*

31. $45 \text{ \sout{egg cartons}} \times \frac{12 \text{ eggs}}{1 \text{ \sout{egg cartons}}} = 540 \text{ eggs can be shipped}$

Although there are more than 540 eggs, the number of egg cartons limits the amount of eggs that can be shipped. Thus, 45 cartons of eggs can be shipped.

33. $2\ SO_{2(g)} + O_{2(g)} \rightarrow 2\ SO_{3(g)}$

$$25.0 \text{ L } SO_2 \times \frac{2 \text{ L } SO_3}{2 \text{ L } SO_2} = 25.0 \text{ L } SO_3$$

$$25.0 \text{ L } O_2 \times \frac{2 \text{ L } SO_3}{1 \text{ L } O_2} = 50.0 \text{ L } SO_3$$

(a) 25.0 L $SO_3$
(b) $SO_2$ is the limiting reactant

35. $2\ Al(OH)_{3(s)} + 3\ H_2SO_{4(aq)} \rightarrow Al_2(SO_4)_{3(aq)} + 6\ H_2O_{(l)}$

$$1.00 \text{ g } Al(OH)_3 \times \frac{1 \text{ mol } Al(OH)_3}{78.0 \text{ g } Al(OH)_3} \times \frac{6 \text{ mol } H_2O}{2 \text{ mol } Al(OH)_3} \times \frac{18.0 \text{ g } H_2O}{1 \text{ mol } H_2O} = 0.692 \text{ g } H_2O$$

$$25.0 \text{ mL } H_2SO_4 \times \frac{0.500 \text{ mol } H_2SO_4}{1000 \text{ mL } H_2SO_4} \times \frac{6 \text{ mol } H_2O}{3 \text{ mol } H_2SO_4} \times \frac{18.0 \text{ g } H_2O}{1 \text{ mol } H_2O} = 0.450 \text{ g } H_2O$$

(a) 0.450 g $H_2O$
(b) $H_2SO_4$ is the limiting reactant

**Section 18.5** *Thermochemical Stoichiometry*

37. For endothermic reactions the $\Delta H$ value is positive because heat is gained.

39. (a) $NH_4NO_{3(s)} \rightarrow N_2O_{(g)} + 2\ H_2O_{(g)} + 47.8 \text{ kcal}$

(b) $6\ CO_{2(g)} + 6\ H_2O_{(l)} + 674.0 \text{ kcal} \rightarrow C_6H_{12}O_{6(s)} + 6\ O_{2(g)}$

41. $C_2H_5OH(g) + 3\ O_2(g) \rightarrow 2\ CO_2(g) + 3\ H_2O(g) + 66.4$ kcal

(a) $1000.0\ \cancel{kcal} \times \dfrac{1\ \cancel{mol\ C_2H_5OH}}{66.4\ \cancel{kcal}} \times \dfrac{46.0\ g\ C_2H_5OH}{1\ \cancel{mol\ C_2H_5OH}} = 693\ g\ C_2H_5OH$

(b) $1000.0\ \cancel{kcal} \times \dfrac{2\ \cancel{mol\ CO_2}}{66.4\ \cancel{kcal}} \times \dfrac{22.4\ L\ CO_2}{1\ \cancel{mol\ CO_2}} = 675\ L\ CO_2$ at STP

**Section 18.6** *Multiple-Reaction Stoichiometry*

43. $2\ C(s) + O_2(g) \rightarrow 2\ CO(g)$
$2\ CO(g) + O_2(g) \rightarrow 2\ CO_2(g)$

(a) $25.0\ \cancel{g\ C} \times \dfrac{1\ \cancel{mol\ C}}{12.0\ \cancel{g\ C}} \times \dfrac{2\ \cancel{mol\ CO}}{2\ \cancel{mol\ C}} \times \dfrac{2\ \cancel{mol\ CO_2}}{2\ \cancel{mol\ CO}} \times \dfrac{44.0\ g\ CO_2}{1\ \cancel{mol\ CO_2}}$
$= 91.7\ g\ CO_2$

(b) $25.0\ \cancel{g\ C} \times \dfrac{1\ \cancel{mol\ C}}{12.0\ \cancel{g\ C}} \times \dfrac{2\ \cancel{mol\ CO}}{2\ \cancel{mol\ C}} \times \dfrac{2\ \cancel{mol\ CO_2}}{2\ \cancel{mol\ CO}} \times \dfrac{22.4\ L\ CO_2}{1\ \cancel{mol\ CO_2}}$
$= 46.7\ L\ CO_2$

(c) 1st reaction:
$25.0\ \cancel{g\ C} \times \dfrac{1\ \cancel{mol\ C}}{12.0\ \cancel{g\ C}} \times \dfrac{1\ \cancel{mol\ O_2}}{2\ \cancel{mol\ C}} \times \dfrac{32.0\ g\ O_2}{1\ \cancel{mol\ O_2}} = 33.3\ g\ O_2$

2nd reaction:
$25.0\ \cancel{g\ C} \times \dfrac{1\ \cancel{mol\ C}}{12.0\ \cancel{g\ C}} \times \dfrac{2\ \cancel{mol\ CO}}{2\ \cancel{mol\ C}} \times \dfrac{1\ \cancel{mol\ O_2}}{2\ \cancel{mol\ CO}} \times \dfrac{32.0\ g\ O_2}{1\ \cancel{mol\ O_2}}$
$= 33.3\ g\ O_2$

Total mass of $O_2$ consumed: $33.3\ g + 33.3\ g = 66.6\ g\ O_2$

45.* $S(s) + O_2(g) \rightarrow SO_2(g)$
$2\ SO_2(g) + O_2(g) \rightarrow 2\ SO_3(g)$
$SO_3(g) + H_2O(l) \rightarrow H_2SO_4(l)$

(a) $1.00\ \cancel{kg\ S} \times \dfrac{1000\ \cancel{g\ S}}{1\ \cancel{kg\ S}} \times \dfrac{1\ \cancel{mol\ S}}{32.1\ \cancel{g\ S}} \times \dfrac{1\ \cancel{mol\ SO_2}}{1\ \cancel{mol\ S}} \times \dfrac{2\ \cancel{mol\ SO_3}}{2\ \cancel{mol\ SO_2}} \times$
$\dfrac{80.1\ \cancel{g\ SO_3}}{1\ \cancel{mol\ SO_3}} \times \dfrac{1\ kg\ SO_3}{1000\ \cancel{g\ SO_3}} = 2.50\ kg\ SO_3$

(b) $1.00\ \cancel{kg\ S} \times \dfrac{1000\ \cancel{g\ S}}{1\ \cancel{kg\ S}} \times \dfrac{1\ \cancel{mol\ S}}{32.1\ \cancel{g\ S}} \times \dfrac{1\ \cancel{mol\ SO_2}}{1\ \cancel{mol\ S}} \times \dfrac{2\ \cancel{mol\ SO_3}}{2\ \cancel{mol\ SO_2}} \times$
$\dfrac{22.4\ L\ SO_3}{1\ \cancel{mol\ SO_3}} = 698\ L\ SO_3$

---

* continued on next page

45. (c) $1.00\ \cancel{\text{kg S}} \times \dfrac{1000\ \cancel{\text{g S}}}{1\ \cancel{\text{kg S}}} \times \dfrac{1\ \cancel{\text{mol S}}}{32.1\ \cancel{\text{g S}}} \times \dfrac{1\ \cancel{\text{mol SO}_2}}{1\ \cancel{\text{mol S}}} \times \dfrac{2\ \cancel{\text{mol SO}_3}}{2\ \cancel{\text{mol SO}_2}} \times$

$$\frac{1\ \cancel{\text{mol H}_2\text{SO}_4}}{1\ \cancel{\text{mol SO}_3}} \times \frac{98.1\ \cancel{\text{g H}_2\text{SO}_4}}{1\ \cancel{\text{mol H}_2\text{SO}_4}} \times \frac{55.0\ \cancel{\text{g H}_2\text{SO}_4}}{100\ \cancel{\text{g H}_2\text{SO}_4}} \times \frac{1\ \text{kg H}_2\text{SO}_4}{1000\ \cancel{\text{g H}_2\text{SO}_4}} = 1.68\ \text{kg H}_2\text{SO}_4$$

47. $2\ NH_{3(aq)} + NaOCl_{(aq)} \rightarrow N_2H_{4(l)} + NaCl_{(aq)} + H_2O_{(l)}$

$2\ N_2H_{4(l)} + N_2O_{4(l)} \rightarrow 3\ N_{2(g)} + 4\ H_2O_{(g)}$

(a) $50.0\ \cancel{\text{mL NH}_3} \times \dfrac{6.00\ \cancel{\text{mol NH}_3}}{1000\ \cancel{\text{mL NH}_3}} \times \dfrac{1\ \cancel{\text{mol N}_2\text{H}_4}}{2\ \cancel{\text{mol NH}_3}} \times \dfrac{3\ \cancel{\text{mol N}_2}}{2\ \cancel{\text{mol N}_2\text{H}_4}} \times \dfrac{28.0\ \text{g N}_2}{1\ \cancel{\text{mol N}_2}} = 6.30\ \text{g N}_2$

(b) $50.0\ \cancel{\text{mL NH}_3} \times \dfrac{6.00\ \cancel{\text{mol NH}_3}}{1000\ \cancel{\text{mL NH}_3}} \times \dfrac{1\ \cancel{\text{mol N}_2\text{H}_4}}{2\ \cancel{\text{mol NH}_3}} \times \dfrac{3\ \cancel{\text{mol N}_2}}{2\ \cancel{\text{mol N}_2\text{H}_4}} \times \dfrac{22.4\ \text{L N}_2}{1\ \cancel{\text{mol N}_2}} = 5.04\ \text{L N}_2$

(c) $50.0\ \cancel{\text{mL NH}_3} \times \dfrac{6.00\ \cancel{\text{mol NH}_3}}{1000\ \cancel{\text{mL NH}_3}} \times \dfrac{1\ \cancel{\text{mol N}_2\text{H}_4}}{2\ \cancel{\text{mol NH}_3}} \times \dfrac{4\ \cancel{\text{mol H}_2\text{O}}}{2\ \cancel{\text{mol N}_2\text{H}_4}} \times \dfrac{18.0\ \text{g H}_2\text{O}}{1\ \cancel{\text{mol H}_2\text{O}}} = 5.40\ \text{g H}_2\text{O}$

(d) $50.0\ \cancel{\text{mL NH}_3} \times \dfrac{6.00\ \cancel{\text{mol NH}_3}}{1000\ \cancel{\text{mL NH}_3}} \times \dfrac{1\ \cancel{\text{mol N}_2\text{H}_4}}{2\ \cancel{\text{mol NH}_3}} \times \dfrac{4\ \cancel{\text{mol H}_2\text{O}}}{2\ \cancel{\text{mol N}_2\text{H}_4}} \times \dfrac{22.4\ \text{L H}_2\text{O}}{1\ \cancel{\text{mol H}_2\text{O}}} = 6.72\ \text{L H}_2\text{O}$

## **Section 18.7** *Advanced Problem-Solving Examples*

49. $n = \dfrac{P\,V}{R\,T}$

$P = 0.974\ \text{atm}$
$V = 2.50\ \text{L}$
$T = 35°\text{C} + 273 = 308\ \text{K}$

$$n = \frac{0.974\ \cancel{\text{atm}} \times 2.50\ \cancel{\text{L}}}{308\ \cancel{\text{K}}} \times \frac{1\ \text{mol} \cdot \cancel{\text{K}}}{0.0821\ \cancel{\text{atm}} \cdot \cancel{\text{L}}} = 0.0963\ \text{mol NO}_2$$

$$0.0963\ \cancel{\text{mol NO}_2} \times \frac{46.0\ \text{g NO}_2}{1\ \cancel{\text{mol NO}_2}} = 4.43\ \text{g NO}_2$$

51. $CaCl_{2(aq)} \rightarrow Ca^{2+}{}_{(aq)} + 2\ Cl^{-}{}_{(aq)}$

$$34.5\ \cancel{g\ CaCl_2} \times \frac{1\ \cancel{mol\ CaCl_2}}{111.1\ \cancel{g\ CaCl_2}} \times \frac{1\ mol\ Ca^{2+}}{1\ \cancel{mol\ CaCl_2}} = 0.311\ mol\ Ca^{2+}$$

$$34.5\ \cancel{g\ CaCl_2} \times \frac{1\ \cancel{mol\ CaCl_2}}{111.1\ \cancel{g\ CaCl_2}} \times \frac{2\ mol\ Cl^{-}}{1\ \cancel{mol\ CaCl_2}} = 0.621\ mol\ Cl^{-}$$

$$\text{Molarity of } Ca^{2+}\text{:}\ \frac{0.311\ mol\ Ca^{2+}}{500.0\ \cancel{mL\ solution}} \times \frac{1000\ \cancel{mL\ solution}}{1\ L\ solution} = 0.622\ M\ Ca^{2+}$$

$$\text{Molarity of } Cl^{-}\text{:}\ \frac{0.621\ mol\ Cl^{-}}{500.0\ \cancel{mL\ solution}} \times \frac{1000\ \cancel{mL\ solution}}{1\ L\ solution} = 1.24\ M\ Cl^{-}$$

53. $2\ HCl_{(aq)} + Zn_{(s)} \rightarrow H_{2(g)} + ZnCl_{2(aq)}$

$$50.0\ \cancel{mL\ H_2} \times \frac{1\ \cancel{L\ H_2}}{1000\ \cancel{mL\ H_2}} \times \frac{1\ \cancel{mol\ H_2}}{22.4\ \cancel{L\ H_2}} \times \frac{2\ \cancel{mol\ HCl}}{1\ \cancel{mol\ H_2}} \times \frac{1000\ mL\ solution}{0.100\ \cancel{mol\ HCl}} = 44.6\ mL\ HCl\ solution$$

55. $n = \frac{PV}{RT}$

$$P = 749\ \cancel{mm\ Hg} \times \frac{1.00\ atm}{760\ \cancel{mm\ Hg}} = 0.986\ atm$$

$V = 0.150\ L$

$T = 25°C + 273 = 298\ K$

$$n = \frac{0.986\ \cancel{atm} \times 0.150\ \cancel{L}}{298\ \cancel{K}} \times \frac{mol \cdot \cancel{K}}{0.0821\ \cancel{atm} \cdot \cancel{L}} = 0.00605\ mol\ N_2O$$

$$0.00605\ \cancel{mol\ N_2O} \times \frac{6.02 \times 10^{23}\ molecules\ N_2O}{1\ \cancel{mol\ N_2O}} = 3.64 \times 10^{21}\ molecules\ N_2O$$

$$0.00605\ \cancel{mol\ N_2O} \times \frac{44.0\ g\ N_2O}{1\ \cancel{mol\ N_2O}} = 0.266\ g\ N_2O$$

## General Exercises

57.*

| | Term | | Term |
|---|---|---|---|
| (a) | density | (b) | specific heat |
| (c) | kilocalorie | (d) | joule |
| (e) | Avogadro's number | (f) | mole |

---

* continued on next page

57.

| | Term | | Term |
|---|---|---|---|
| (g) | molar mass | (h) | molar volume |
| (i) | empirical formula | (j) | molecular formula |
| (k) | Conservation of Mass law | (l) | stoichiometry |
| (m) | combined gas law | (n) | ideal gas law |
| (o) | gas constant | (p) | mass/mass % concentration |
| (q) | molarity | (r) | molality |

59. $$1.4 \times 10^{21}\ \cancel{\text{L seawater}} \times \frac{1000\ \cancel{\text{mL seawater}}}{1\ \cancel{\text{L seawater}}} \times \frac{1.05\ \cancel{\text{g seawater}}}{1\ \cancel{\text{mL seawater}}} \times \frac{1.90\ \text{g Cl}^-}{100\ \cancel{\text{g seawater}}} = 2.8 \times 10^{22}\ \text{g chlorine}$$

61. <u>Empirical Formula of Mannitol</u>

$$39.6\ \cancel{\text{g C}} \times \frac{1\ \text{mol C}}{12.0\ \cancel{\text{g C}}} = 3.30\ \text{mol C}$$

$$7.7\ \cancel{\text{g H}} \times \frac{1\ \text{mol H}}{1.0\ \cancel{\text{g H}}} = 7.7\ \text{mol H}$$

$$52.8\ \cancel{\text{g O}} \times \frac{1\ \text{mol O}}{16.0\ \cancel{\text{g O}}} = 3.30\ \text{mol O}$$

$$\frac{C_{3.30}}{3.30}\frac{H_{7.7}}{3.30}\frac{O_{3.30}}{3.30} = C_{1.00}H_{2.3}O_{1.00}$$

Empirical formula: $3\,(C_{1.00}H_{2.3}O_{1.00}) = C_3H_7O_3$

MM of $C_3H_7O_3$ = 3(12.0) g + 7(1.0) g + 3(16.0) g = 91.0 g/mol

$$\text{MM of Mannitol} = \frac{1\ \cancel{\text{mL}}}{4.93 \times 10^{21}\ \cancel{\text{molecules}}} \times \frac{1.47\ \text{g}}{1\ \cancel{\text{mL}}} \times \frac{6.02 \times 10^{23}\ \cancel{\text{molecules}}}{1\ \text{mol}} = 180\ \text{g/mol}$$

<u>Molecular Formula of Mannitol</u>

$$\frac{180\ \text{g/mol}}{91.0\ \text{g/mol}} \approx 2$$

Molecular formula: $2\,(C_3H_7O_3) = C_6H_{14}O_6$

63. $$50.0\ \cancel{\text{mL NaCl solution}} \times \frac{0.100\ \cancel{\text{mol NaCl}}}{1000\ \cancel{\text{mL NaCl solution}}} \times \frac{1\ \text{mol Na}^+}{1\ \cancel{\text{mol NaCl}}}$$
$$= 0.00500\ \text{mol Na}^+ \text{ from NaCl}$$

$$50.0\ \cancel{\text{mL Na}_2\text{SO}_4\text{ solution}} \times \frac{0.200\ \cancel{\text{mol Na}_2\text{SO}_4}}{1000\ \cancel{\text{mL Na}_2\text{SO}_4\text{ solution}}} \times \frac{2\ \text{mol Na}^+}{1\ \cancel{\text{mol Na}_2\text{SO}_4}}$$
$$= 0.0200\ \text{mol Na}^+ \text{ from Na}_2\text{SO}_4$$

Total mol of $Na^+$: 0.00500 mol + 0.0200 mol = 0.0250 mol $Na^+$

Concentration of $Na^+$:
$$\frac{0.0250\ \text{mol Na}^+}{(50.0 + 50.0)\ \cancel{\text{mL solution}}} \times \frac{1000\ \cancel{\text{mL solution}}}{1\ \text{L solution}} = 0.250\ M\ \text{Na}^+$$

65. $Na_2CO_{3(s)} + 2\ HCl_{(aq)} \rightarrow 2\ NaCl_{(aq)} + H_2O_{(l)} + CO_{2(g)}$
$HCl_{(aq)} + NaOH_{(aq)} \rightarrow NaCl_{(aq)} + H_2O_{(l)}$

Total amount of HCl:
$$50.0\ \cancel{\text{mL HCl}} \times \frac{0.345\ \text{mol HCl}}{1000\ \cancel{\text{mL HCl}}} = 0.0173\ \text{mol HCl}$$

Amount of HCl neutralized by NaOH:
$$15.9\ \cancel{\text{mL NaOH}} \times \frac{0.155\ \cancel{\text{mol NaOH}}}{1000\ \cancel{\text{mL NaOH}}} \times \frac{1\ \text{mol HCl}}{1\ \cancel{\text{mol NaOH}}} = 0.00246\ \text{mol HCl}$$

Amount of HCl neutralized by $Na_2CO_3$:
0.0173 mol – 0.00246 mol = 0.01484 mol HCl

Mass of $Na_2CO_3$:
$$0.01484\ \cancel{\text{mL HCl}} \times \frac{1\ \cancel{\text{mol Na}_2\text{CO}_3}}{2\ \cancel{\text{mL HCl}}} \times \frac{106.0\ \text{g Na}_2\text{CO}_3}{1\ \cancel{\text{mol Na}_2\text{CO}_3}} = 0.787\ \text{g Na}_2\text{CO}_3$$

CHAPTER

# 19 Nuclear Chemistry

## Section 19.1 *Natural Radioactivity*

1. <u>Principal Types of Natural Radiation</u>
   (1) alpha
   (2) beta
   (3) gamma

3. An alpha particle ($\alpha$) is identical to a helium-4 nucleus.

5. A beta particle ($\beta$) is identical to an electron.

7. Alpha particles ($\alpha$) require heavy paper or clothing as minimum shielding and cannot penetrate human skin.

9. Beta particles ($\beta$) are emitted from the nucleus at ~ 90% the velocity of light.

11. Gamma rays ($\gamma$) are emitted from the nucleus at the velocity of light.

## Section 19.2 *Nuclear Equations*

13.

| | | Notation | | | Notation |
|---|---|---|---|---|---|
| (a) | alpha particle | ${}^{4}_{2}He$ | (b) | beta particle | ${}^{0}_{-1}e$ |
| (c) | gamma ray | ${}^{0}_{0}\gamma$ | (d) | positron | ${}^{0}_{+1}e$ |
| (e) | neutron | ${}^{1}_{0}n$ | (f) | proton | ${}^{1}_{1}H$ |

15. (a) ${}^{175}_{78}Pt \rightarrow {}^{171}_{76}Os + {}^{4}_{2}He$ (b) ${}^{28}_{13}Al \rightarrow {}^{28}_{14}Si + {}^{0}_{-1}e$

(c) ${}^{55}_{27}Co \rightarrow {}^{55}_{26}Fe + {}^{0}_{+1}e$ (d) ${}^{44}_{22}Ti + {}^{0}_{-1}e \rightarrow {}^{44}_{21}Sc$

17.

(a) ${}^{221}_{88}X \rightarrow {}^{217}_{86}Rn + {}^{4}_{2}He$

(b) ${}^{43}_{19}X \rightarrow {}^{43}_{20}Ca + {}^{0}_{-1}e$

(c) ${}^{19}_{10}X \rightarrow {}^{19}_{9}F + {}^{0}_{+1}e$

(d) ${}^{37}_{18}X + {}^{0}_{-1}e \rightarrow {}^{37}_{17}Cl$

Answer

${}^{221}_{88}X = {}^{221}_{88}Ra$

${}^{43}_{19}X = {}^{43}_{19}K$

${}^{19}_{10}X = {}^{19}_{10}Ne$

${}^{37}_{18}X = {}^{37}_{18}Ar$

**Section 19.3** *Radioactive Decay Series*

19. Let X = the parent isotope
${}^{210}_{84}X \rightarrow {}^{206}_{82}Pb + {}^{4}_{2}He$ Parent isotope: ${}^{210}_{84}Po$

21. ${}^{207}_{81}X \rightarrow {}^{207}_{82}Pb + {}^{0}_{-1}e$ Parent isotope: ${}^{207}_{81}Tl$

23. ${}^{212}_{84}X \rightarrow {}^{208}_{82}Pb + {}^{4}_{2}He$ Parent isotope: ${}^{212}_{84}Po$

25. Decay Series for Thorium-232

${}^{232}_{90}Th \xrightarrow{\alpha} {}^{228}_{88}Ra \xrightarrow{\beta} {}^{232}_{89}Ac$

$\downarrow \beta$

${}^{220}_{86}Rn \xleftarrow{\alpha} {}^{224}_{88}Ra \xleftarrow{\alpha} {}^{228}_{90}Th$

$\alpha \downarrow$

${}^{216}_{84}Po \xrightarrow{\alpha} {}^{212}_{82}Pb \xrightarrow{\beta} {}^{212}_{83}Bi$

$\downarrow \beta$

${}^{208}_{82}Pb \xleftarrow{\alpha} {}^{212}_{84}Po$

Answers

| | | | |
|---|---|---|---|
| (a) | $\alpha$ | (b) | $\beta$ |
| (c) | $\beta$ | (d) | $\alpha$ |
| (e) | $\alpha$ | (f) | ${}^{216}_{84}Po$ |
| (g) | ${}^{212}_{82}Pb$ | (h) | ${}^{212}_{83}Bi$ |
| (i) | ${}^{212}_{84}Po$ | (j) | ${}^{208}_{82}Pb$ |

**Section 19.4** *Radioactive Half-Life*

27. The decay of a nucleus by particle emission is recorded when a Geiger counter clicks.

29. Half of 100% of a radioisotope is 50%. Half of 50% of a radioisotope is 25%. Thus, it takes two half-lives to reach 25% of an original radioisotope sample from the complete sample.

31. $15\ t_{1/2} \times \dfrac{24{,}400 \text{ years}}{1\ t_{1/2}} = 366{,}000 \text{ years}$

33.

| Activity(%) | Elapsed Time ($t_{1/2}$) |
|---|---|
| 100 | 0 |
| 50 | 1 |
| 25 | 2 |
| 12.5 | 3 |
| 6.25 | 4 |

$$4\ t_{1/2} \times \frac{5730 \text{ years}}{1\ t_{1/2}} \approx 22{,}900 \text{ years old}$$

35. $$60 \text{ hours} \times \frac{1\ t_{1/2}}{15 \text{ hours}} = 4\ t_{1/2}$$

$$80 \text{ mg Na-24} \times \frac{1}{2} \times \frac{1}{2} \times \frac{1}{2} \times \frac{1}{2} = 5 \text{ mg Na-24}$$

37.

| Activity(dpm) | Elapsed Time ($t_{1/2}$) |
|---|---|
| 560 | 0 |
| 280 | 1 |
| 140 | 2 |
| 70 | 3 |
| 35 | 4 |

$$4\ t_{1/2} \times \frac{74 \text{ days}}{1\ t_{1/2}} = 296 \text{ days}$$

39.

| Activity(dpm) | Elapsed Time ($t_{1/2}$) |
|---|---|
| 1200 | 0 |
| 600 | 1 |
| 300 | 2 |
| 150 | 3 |
| 75 | 4 |

$$\text{half-life} \times 4\ t_{1/2} = 21.2 \text{ years}$$

$$\text{half-life} = \frac{21.2 \text{ years}}{4\ t_{1/2}} = \frac{5.30 \text{ years}}{1\ t_{1/2}}$$

**Section 19.5** *Applications of Radioisotopes*

41. Fossils up to 50,000 years old can be dated using radiocarbon dating (carbon-14).

43. Uranium-lead dating can be used to estimate geological events occurring billions of years ago.

45. The β-emitting radioisotope iodine-131 can be used to measure the activity of the thyroid gland.

47. The γ-emitting radioisotope iridium-192 can be inserted directly into breast tissue for the treatment of tumors.

49. The γ-emitting radioisotope technetium-99 can be used to diagnose and locate brain tumors.

51. Plutonium-238 releases high-energy gamma rays that are used to power heart pacemakers.

**Section 19.6** *Artificial Radioactivity*

53. $^{14}_{7}N + ^{4}_{2}He \rightarrow ^{1}_{1}H + ^{17}_{8}O$ Rutherford obtained: $^{17}_{8}O$

55. $^{9}_{4}Be + ^{4}_{2}He \rightarrow ^{1}_{0}n + ^{12}_{6}C$ Chadwick obtained: $^{12}_{6}C$

57. $^{27}_{13}Al + ^{4}_{2}He \rightarrow ^{1}_{0}n + ^{30}_{15}P$ Curie and Joliot obtained: $^{30}_{15}P$

59. $^{10}_{5}B + ^{4}_{2}He \rightarrow ^{14}_{7}N + ^{0}_{0}\gamma$ The projectile particle is: $^{4}_{2}He$

61. $^{26}_{12}Mg + ^{2}_{1}H \rightarrow ^{27}_{12}Mg + ^{1}_{1}H$ The projectile particle is: $^{2}_{1}H$

63. $^{12}_{6}C + ^{1}_{1}H \rightarrow ^{13}_{7}N + ^{0}_{0}\gamma$ The projectile particle is: $^{1}_{1}H$

65. The Russians created: $^{260}_{105}Ns$

67. $^{54}_{24}Cr + ^{207}_{82}Pb \rightarrow ^{1}_{0}n + ^{260}_{106}Unh$ The Russians created: $^{260}_{106}Unh$

69. $^{142}_{60}Nd + ^{1}_{0}n \rightarrow ^{0}_{-1}e + ^{143}_{61}Pm$

**Section 19.7** *Nuclear Fission*

71. $^{1}_{0}n \xrightarrow{\text{1st}} 2\,^{1}_{0}n \xrightarrow{\text{2nd}} 4\,^{1}_{0}n \xrightarrow{\text{3rd}} 8\,^{1}_{0}n$
After the third step, 8 neutrons are produced.

73. Uranium–235 can fission in over 40 ways to give different products. The number of neutrons released from a single fission is usually 2 or 3, but the *average* number of neutrons is 2.5.

75. $^{235}_{92}U + ^{1}_{0}n \rightarrow ^{144}_{54}Xe + ^{90}_{38}Sr + 2\,^{1}_{0}n$
This fission reaction produces two neutrons.

77. $^{239}_{94}Pu + ^{1}_{0}n \rightarrow ^{137}_{55}Cs + ^{100}_{39}Y + 3\,^{1}_{0}n$
This fission reaction produces three neutrons.

79. $^{233}_{92}U + ^{1}_{0}n \rightarrow ^{143}_{56}Ba + ^{88}_{36}Kr + 3\,^{1}_{0}n$
Uranium-233 undergoes fission to give barium-143, krypton-88, and two neutrons.

**Section 19.8** *Mass Defect and Binding Energy*

81. $^{4}_{2}He$ has two protons and two neutrons
2(1.0073 amu) + 2(1.0087 amu) = 4.0320 amu

Mass defect: 4.0320 amu – 4.0015 amu = 0.0305 amu

83. Helium-3 has two protons and one neutron
2(1.0073 amu) + (1.0087 amu) = 3.0233 amu

Mass defect: 3.0233 amu – 3.0149 amu = 0.0084 amu

85. Lithium-7 has three protons and four neutrons
3(1.0073 amu) + 4(1.0087 amu) = 7.0567 amu

Mass defect: 7.0567 amu – 7.0160 amu = 0.0407 amu

Binding energy: $E = mc^2$

$$E = \frac{0.0407\ g}{1\ mol} \times \frac{1\ kg}{1000\ g} \times \left(\frac{3.00 \times 10^8\ m}{s}\right)^2 = 3.66 \times 10^{12}\ J/mol$$

87. Boron-11 has five protons and six neutrons
5(1.0073 amu) + 6(1.0087 amu) = 11.0887 amu

Mass defect: 11.0887 amu – 11.0093 amu = 0.0794 amu

Binding energy: $E = mc^2$

$$E = \frac{0.0794\ g}{1\ mol} \times \frac{1\ kg}{1000\ g} \times \left(\frac{3.00 \times 10^8\ m}{s}\right)^2 = 7.15 \times 10^{12}\ J/mol$$

89. Nitrogen-14 has seven protons and seven neutrons
$7(1.0073 \text{ amu}) + 7(1.0087 \text{ amu}) = 14.1120 \text{ amu}$

Mass defect: $14.1120 \text{ amu} - 14.0031 \text{ amu} = 0.1089 \text{ amu}$

Binding energy: $E = mc^2$

$$E = \frac{0.1089 \text{ g}}{1 \text{ mol}} \times \frac{1 \text{ kg}}{1000 \text{ g}} \times \left(\frac{3.00 \times 10^8 \text{ m}}{\text{s}}\right)^2 = 9.80 \times 10^{12} \text{ J/mol}$$

91. Fluorine-20 has nine protons and eleven neutrons
$9(1.0073 \text{ amu}) + 11(1.0087 \text{ amu}) = 20.1614 \text{ amu}$

Mass defect: $20.1614 \text{ amu} - 19.9999 \text{ amu} = 0.1615 \text{ amu}$

Binding energy: $E = mc^2$

$$E = \frac{0.1615 \text{ g}}{1 \text{ mol}} \times \frac{1 \text{ kg}}{1000 \text{ g}} \times \left(\frac{3.00 \times 10^8 \text{ m}}{\text{s}}\right)^2 = 1.45 \times 10^{13} \text{ J/mol}$$

## Section 19.9 *Nuclear Fusion*

93. $^{1}_{1}\text{H} + {}^{1}_{1}\text{H} \rightarrow {}^{2}_{1}\text{H} + {}^{0}_{+1}\text{X}$ Particle X: $^{0}_{+1}\text{e}$

95. $^{2}_{1}\text{H} + {}^{2}_{1}\text{H} \rightarrow {}^{3}_{1}\text{H} + {}^{0}_{+1}\text{e} + {}^{1}_{0}\text{X}$ Particle X: $^{1}_{0}\text{n}$

97. $^{3}_{2}\text{He} + {}^{3}_{2}\text{He} \rightarrow 2\,{}^{1}_{1}\text{H} + {}^{4}_{2}\text{X}$ Particle X: $^{4}_{2}\text{He}$

## Section 19.10 *Nuclear Reactors and Power Plants*

99. The three radioisotopes used in the fuel rods of a nuclear reactor core are $^{235}_{92}\text{U}$, $^{233}_{92}\text{U}$, and $^{239}_{94}\text{Pu}$.

101. The control rods of the reactor core regulate the rate of fission by absorbing neutrons.

103. $^{238}_{92}\text{U} + {}^{1}_{0}\text{n} \rightarrow {}^{239}_{92}\text{U}$
$^{239}_{92}\text{U} \rightarrow {}^{239}_{93}\text{Np} + {}^{0}_{-1}\text{e}$
$^{239}_{93}\text{Np} \rightarrow {}^{239}_{94}\text{Pu} + {}^{0}_{-1}\text{e}$ The resulting fissionable isotope is: $^{239}_{94}\text{Pu}$

105. A "light water" reactor uses $H_2O$ as the core coolant.
A "heavy water" reactor uses $D_2O$, that is heavy water, as the coolant.

## General Exercises

107. (a) $^{241}_{94}Pu \rightarrow ^{4}_{2}He + ^{237}_{92}U$ The daughter product is: $^{237}_{92}U$

(b) $^{237}_{92}U \rightarrow ^{0}_{-1}e + ^{237}_{93}Np$ The granddaughter product is: $^{237}_{93}Np$

109. $^{55}_{26}Fe + ^{0}_{-1}e \rightarrow ^{55}_{25}Mn + ^{0}_{0}\gamma$

111. Since an activity of ~ 7 dpm per gram indicates that one half-life has expired, the age of Crater Lake is ~6000 years because $1\ t_{1/2} = 5730$ years.

113. $^{0}_{+1}e + ^{0}_{-1}e \rightarrow 2\ ^{0}_{0}\gamma$

115. $^{238}_{92}U + 17\ ^{1}_{0}n \rightarrow ^{255}_{100}Fm + 8\ ^{0}_{-1}e$

Notice that 17 neutrons are needed to balance this nuclear equation.

117. $^{58}_{26}Fe + ^{208}_{82}Pb \rightarrow ^{1}_{0}n + ^{265}_{108}Uno$ The Germans produced: $^{265}_{108}Uno$

119. $^{6}_{3}Li + ^{1}_{0}n \rightarrow ^{3}_{1}H + ^{4}_{2}X$ + energy Particle X: $^{4}_{2}He$

# Organic Chemistry

CHAPTER 20

## Section 20.1 *Hydrocarbons*

1. ~ 90% of all chemical compounds are organic.

3. The primary source of hydrocarbons is petroleum (crude oil).

5.

| | Hydrocarbon | Classification |
|---|---|---|
| (a) | alkanes | saturated |
| (b) | alkenes | unsaturated |
| (c) | alkynes | unsaturated |

## Section 20.1 *Alkanes*

7. General Molecular Formula for Alkanes

$C_nH_{2n+2}$

9.

| | Alkane | Name |
|---|---|---|
| (a) | $CH_3CH_2CH_2CH_2CH_2CH_3$ | hexane |
| (b) | $CH_3CH_2CH_2CH_3$ | butane |
| (c) | $CH_3(CH_2)_6CH_3$ | octane |
| (d) | $CH_3(CH_2)_8CH_3$ | decane |

11. Three Isomers of Pentane ($C_5H_{12}$)

$CH_3 - CH_2 - CH_2 - CH_2 - CH_3$

$$\begin{array}{c} \phantom{CH_3 -{}} CH_3 \phantom{{}- CH_2 - CH_3} \\ \phantom{CH_3 -{}} | \phantom{{}- CH_2 - CH_3} \\ CH_3 - CH - CH_2 - CH_3 \end{array}$$

$$\begin{array}{c} CH_3 \\ | \\ CH_3 - C - CH_3 \\ | \\ CH_3 \end{array}$$

13. Two Isomers of Bromopropane ($C_3H_7Br$)

$$\begin{array}{ccccccc} & H & & H & & H & \\ & | & & | & & | & \\ H- & C & - & C & - & C & -Br \\ & | & & | & & | & \\ & H & & H & & H & \end{array} \quad \text{and} \quad \begin{array}{ccccccc} & H & & Br & & H & \\ & | & & | & & | & \\ H- & C & - & C & - & C & -H \\ & | & & | & & | & \\ & H & & H & & H & \end{array}$$

15. Differences in Properties of Structural Isomers
Melting point, boiling point, density, solubility, etc...

17.

| | Alkyl Substituents | Name |
|---|---|---|
| (a) | $CH_3$– | methyl |
| (b) | $CH_3CH_2$– | ethyl |
| (c) | $CH_3CH_2CH_2$– | propyl |
| (d) | $CH_3CH_2CH_2CH_2$– | butyl |

19. (a) 2-methylpentane

$$\begin{array}{l} \qquad\qquad\qquad\qquad\qquad CH_3 \\ \qquad\qquad\qquad\qquad\qquad | \\ CH_3-CH_2-CH_2-CH-CH_3 \end{array}$$

(b) 3,3,5-trimethylheptane
(Note: The longest continuous carbon chain is 7 atoms.)

$$\begin{array}{l} CH_3-CH_2 \qquad\quad CH_3 \\ \qquad\quad | \qquad\qquad\quad | \\ CH_3-C-CH_2-CH-CH_2-CH_3 \\ \qquad\quad | \\ \qquad\quad CH_3 \end{array}$$

(c) 2,4,4-trimethylheptane

$$\begin{array}{l} \qquad\qquad\qquad\qquad\quad CH_3 \\ \qquad\qquad\qquad\qquad\quad | \\ CH_3-CH-CH_2-C-CH_2-CH_2-CH_3 \\ \qquad\quad | \qquad\qquad\quad | \\ \qquad\quad CH_3 \qquad\quad CH_3 \end{array}$$

(d) 3,4,5-trimethyloctane

$$\begin{array}{l} \qquad\qquad\qquad\qquad\qquad\qquad\quad CH_3 \\ \qquad\qquad\qquad\qquad\qquad\qquad\quad | \\ CH_3-CH_2-CH_2-CH-CH-CH-CH_2-CH_3 \\ \qquad\qquad\qquad\qquad\quad | \qquad\qquad\quad | \\ \qquad\qquad\qquad\qquad\quad CH_3 \qquad\quad CH_3 \end{array}$$

21. When a hydrocarbon undergoes complete oxidation, the two principal products are $CO_2$ and $H_2O$.

23. (a) $CH_4 + 2\,O_2 \rightarrow CO_2 + 2\,H_2O$
(b) $C_5H_{12} + 8\,O_2 \rightarrow 5\,CO_2 + 6\,H_2O$
(c) $2\,C_2H_6 + 7\,O_2 \rightarrow 4\,CO_2 + 6\,H_2O$
(d) $2\,C_6H_{14} + 19\,O_2 \rightarrow 12\,CO_2 + 14\,H_2O$

25. Either a flame or an electrical spark can initiate the combustion of an alkane.

27. (a) $CH_4 + Br_2 \xrightarrow{UV} CH_3Br + HBr$
(b) $C_2H_6 + Br_2 \xrightarrow{UV} C_2H_5Br + HBr$

## Section 20.3 *Alkenes and Alkynes*

29. General Molecular Formula for Alkenes
$C_nH_{2n}$

31.

| | Alkene | Name |
|---|---|---|
| (a) | $CH_3CH{=}CHCH_2CH_2CH_3$ | 2-hexene |
| (b) | $CH_2{=}CHCH_2CH_2CH_3$ | 1-pentene |
| (c) | $CH_2{=}CHCH_2CH_3$ | 1-butene |
| (d) | $CH_3CH_2CH_2CH{=}CHCH_2CH_2CH_3$ | 4-octene |

33. General Molecular Formula for Alkenes
$C_nH_{2n-2}$

35.

| | Alkyne | Name |
|---|---|---|
| (a) | $CH{\equiv}CCH_2CH_2CH_3$ | 1-pentyne |
| (b) | $CH_3C{\equiv}CCH_2CH_3$ | 2-pentyne |
| (c) | $CH_3C{\equiv}CCH_2CH_2CH_3$ | 2-hexyne |
| (d) | $CH_3CH_2C{\equiv}CCH_2CH_2CH_3$ | 3-heptyne |

37. Two Isomers of Pentene ($C_5H_{10}$)
$CH_2 = CH- CH_2 - CH_2 - CH_3$ and $CH_3 - CH = CH - CH_2 - CH_3$

39. Two Geometric Isomers of Pentene ($C_5H_{10}$)

```
CH3     CH2CH3          CH3    H
   \   /                   \  /
    C=C                     C=C
   /   \                   /   \
  H     H                 H     CH2CH3
    cis                      trans
```

41. 

| | Isomer | Physical Properties |
|---|---|---|
| (a) | structural isomers | different |
| (b) | geometric isomers | different |

43. (a) 4-methyl-2-pentene

$$CH_3-CH=CH-\underset{}{\overset{CH_3}{\overset{|}{C}}H}-CH_3$$

(b) 3,5,5-trimethyl-2-heptene
(Note: The longest continuous carbon chain is 7 atoms.)

$$CH_3-\underset{\underset{CH_3}{|}}{\overset{\overset{CH_3-CH_2}{|}}{C}}-CH_2-\overset{\overset{CH_3}{|}}{C}=CH-CH_3$$

(c) 4,4-dimethyl-2-heptyne

$$CH_3-C\equiv C-\underset{\underset{CH_3}{|}}{\overset{\overset{CH_3}{|}}{C}}-CH_2-CH_2-CH_3$$

(d) 3,4,5-trimethyl-1-octyne

$$CH_3-CH_2-CH_2-\underset{\underset{CH_3}{|}}{CH}-\overset{\overset{CH_3}{|}}{CH}-\underset{\underset{CH_3}{|}}{CH}-C\equiv CH$$

45. Ni, Pt, or Pd can act as a catalyst for the hydrogenation of an alkene.

47. (a) $CH_2=CH_2 + 3\,O_2 \rightarrow 2\,CO_2 + 2\,H_2O$

(b) $CH_3CH=CH_2 + Cl_2 \rightarrow CH_3-\underset{\underset{Cl}{|}}{CH}-\underset{\underset{Cl}{|}}{CH_2}$

(c) $CH_3CH=CHCH_3 + Br_2 \rightarrow CH_3-\underset{\underset{Br}{|}}{CH}-\underset{\underset{Br}{|}}{CH}-CH_3$

(d) $CH_2=CHCH_2CH_3 + I_2 \rightarrow \underset{\underset{I}{|}}{CH_2}-\underset{\underset{I}{|}}{CH}-CH_2-CH_3$

## Section 20.4 *Aromatic Hydrocarbons*

49. Two KeKulé Structures of Benzene ($C_6H_6$)

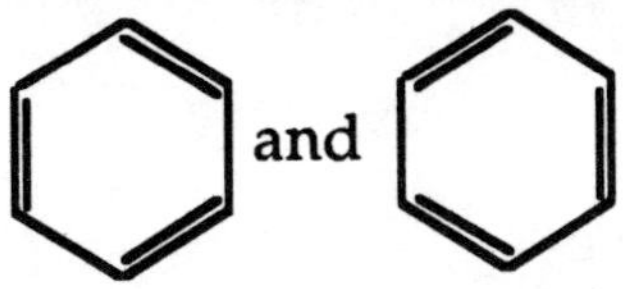

and

51. Three Isomers of Dinitrobenzene [$C_6H_4(NO_2)_2$]

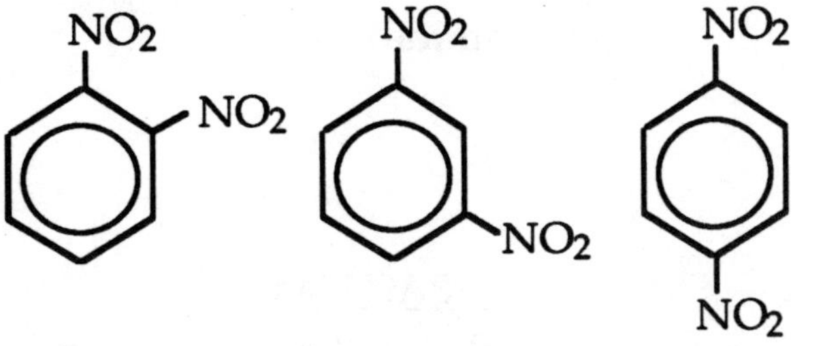

$NO_2$ $NO_2$ $NO_2$

$NO_2$

$NO_2$

$NO_2$

*ortho* *meta* *para*

## Section 20.5 *Hydrocarbon Derivatives*

53.

| | General Formula | Class of Compound |
|---|---|---|
| (a) | R—O—R′ | ethers |
| (b) | R—X | organic halides |
| (c) | Ar—OH | phenols |
| (d) | $R—NH_2$ | amines |
| (e) | R—OH | alcohols |

55.*

| | Chemical Formula | Class of Compound |
|---|---|---|
| (a) | $CH_3–NH_2$ | amine |
| (b) | $CH_3CH_2–F$ | organic halide |
| (c) | $CH_3-\overset{\overset{\large O}{\Vert}}{C}-O-CH_3$ | ester |
| (d) | $H-\overset{\overset{\large O}{\Vert}}{C}-OH$ | carboxylic acid |
| (e) | 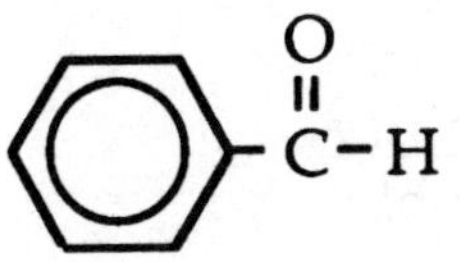 $C_6H_5-\overset{\overset{\large O}{\Vert}}{C}-H$ | aldehyde |

---

* continued on next page

55.

| | Chemical Formula | Class of Compound |
|---|---|---|
| (f) | $CH_3CH_2-O-C_6H_5$ | ether |
| (g) | $C_6H_5-C(=O)-CH_3$ | ketone |
| (h) | $C_6H_5-C(=O)-NH_2$ | amide |

**Section 20.6** *Organic Halides*

57.

| | Organic Halide | IUPAC Name |
|---|---|---|
| (a) | $CH_3I$ | iodomethane |
| (b) | $CH_3CH_2Br$ | bromoethane |
| (c) | $CH_3CH_2CH_2F$ | 1-fluoropropane |
| (d) | $(CH_3)_2CCl_2$ | 2,2-dichloropropane |

59.

| | Organic Halide | Structure |
|---|---|---|
| (a) | methyl chloride | $CH_3-Cl$ |
| (b) | ethyl iodide | $CH_3-CH_2-I$ |
| (c) | 1,2,3-trifluoropropane | $CH_2F-CHF-CH_2F$ |
| (d) | 2,2-dibromopropane | $CH_3-CBr_2-CH_3$ |

61. 1,1,1-trichloroethane

$$Cl-CCl_2-CH_3$$

63.

| | Solvent | Solubility |
|---|---|---|
| (a) | water | insoluble |
| (b) | hydrocarbons | soluble |

## Section 20.7 *Alcohols, Phenols, and Ethers*

65.

| | Alcohol | IUPAC Name |
|---|---|---|
| (a) | $CH_3OH$ | methanol |
| (b) | $CH_3CH_2CH_2OH$ | 1-propanol |
| (c) | $CH_3CH_2OH$ | ethanol |
| (d) | $CH_3CH_2CH(OH)CH_3$ | 2-butanol |

67.

| | Phenol | Name |
|---|---|---|
| (a) | 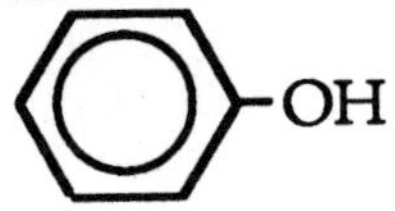 $C_6H_5$–OH | phenol |
| (b) |  $CH_3$–$C_6H_4$–OH | *para*-methylphenol |

69.

| | Ether | Name |
|---|---|---|
| (a) | $CH_3$–O–$CH_3$ | dimethyl ether |
| (b) | $CH_3$–O–$CH_2CH_3$ | methyl ethyl ether |
| (c) | $CH_3CH_2CH_2$–O–$CH_2CH_2CH_3$ | dipropyl ether |
| (d) | 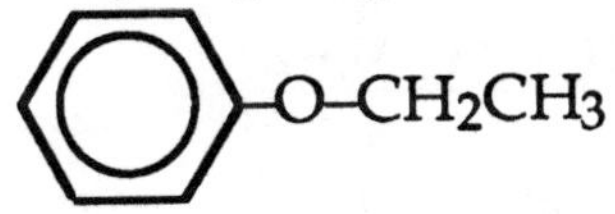 $C_6H_5$–O–$CH_2CH_3$ | phenyl ethyl ether |

71. Hydrogen Bond Between Ethyl Alcohol Molecules

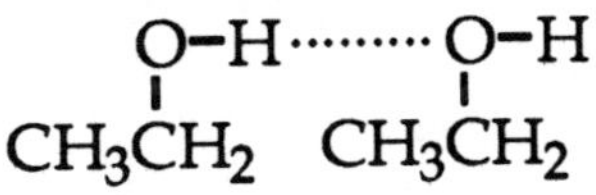

```
   O–H·········O–H
   |           |
CH3CH2      CH3CH2
```

73.

| | | Solubility in Water |
|---|---|---|
| (a) | alcohols | soluble |
| (b) | ethers | insoluble |

75.

| Ether ($C_2H_6O$) | Alcohol ($C_2H_6O$) |
|---|---|
| $CH_3-O-CH_3$ | $CH_3-CH_2-OH$ |

77. $CO_{(g)} + 2\ H_{2(g)} \xrightarrow{\text{Zn/Cr oxides}} CH_3OH_{(g)}$

79. $CH_2{=}CHCH_3 + H_2O \xrightarrow{\text{Acid}} CH_3{-}CH(OH){-}CH_3$

## Section 20.8 *Amines*

81.

| | Amine | Name |
|---|---|---|
| (a) | $CH_3CH_2NH_2$ | ethyl amine |
| (b) | $CH_3CH_2CH_2NH_2$ | propyl amine |
| (c) | $(CH_3)_3N$ | trimethyl amine |
| (d) | 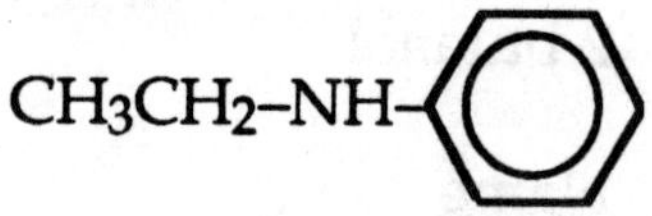 $CH_3CH_2\text{–}NH\text{–}C_6H_5$ | ethyl phenyl amine |

83. Hydrogen Bond Between Ethyl Amine Molecules

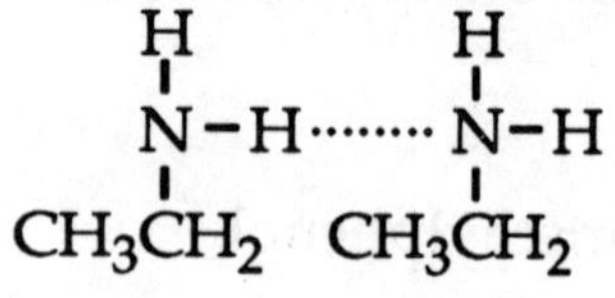

$$\begin{array}{ccc} \text{H} & & \text{H} \\ | & & | \\ \text{N–H} & \cdots\cdots & \text{N–H} \\ | & & | \\ CH_3CH_2 & & CH_3CH_2 \end{array}$$

85. Triethyl amine does not have a H atom attached to the N. Therefore, the N atom has no capacity for hydrogen bonding with water molecules.

## Section 20.9 *Aldehydes and Ketones*

87.

| | Aldehyde | IUPAC Name |
|---|---|---|
| (a) | 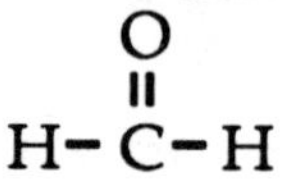 $H\text{–}\overset{\overset{\displaystyle O}{\|\|}}{C}\text{–}H$ | methanal |
| (b) | 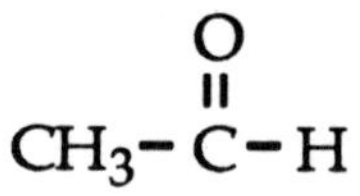 $CH_3\text{–}\overset{\overset{\displaystyle O}{\|\|}}{C}\text{–}H$ | ethanal |
| (c) | 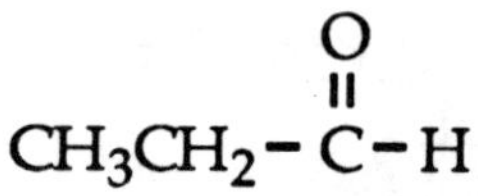 $CH_3CH_2\text{–}\overset{\overset{\displaystyle O}{\|\|}}{C}\text{–}H$ | propanal |
| (d) | 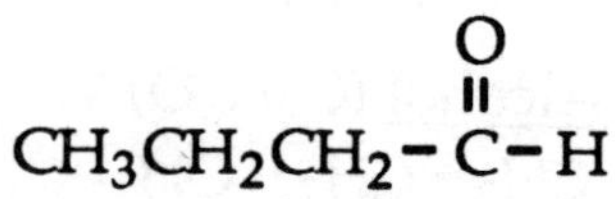 $CH_3CH_2CH_2\text{–}\overset{\overset{\displaystyle O}{\|\|}}{C}\text{–}H$ | butanal |

89.*

| | Ketone | IUPAC Name |
|---|---|---|
| (a) | 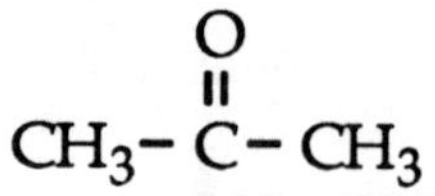 $CH_3\text{–}\overset{\overset{\displaystyle O}{\|\|}}{C}\text{–}CH_3$ | 2-propanone |
| (b) | 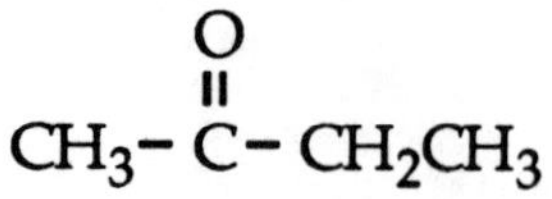 $CH_3\text{–}\overset{\overset{\displaystyle O}{\|\|}}{C}\text{–}CH_2CH_3$ | 2-butanone |

---

* continued on next page

89. Ketone — IUPAC Name

| | Ketone | IUPAC Name |
|---|---|---|
| (c) | $CH_3-\overset{\overset{\displaystyle O}{\|\|}}{C}-CH_2CH_2CH_3$ | 2-pentanone |
| (d) | $CH_3CH_2-\overset{\overset{\displaystyle O}{\|\|}}{C}-CH_2CH_3$ | 3-pentanone |

91.

| | | Higher Boiling Point |
|---|---|---|
| (a) | aldehydes vs. hydrocarbons | aldehydes |
| (b) | ketones vs. hydrocarbons | ketones |

93.

| Aldehyde ($C_3H_6O$) | Ketone ($C_3H_6O$) |
|---|---|
| $CH_3CH_2-\overset{\overset{\displaystyle O}{\|\|}}{C}-H$ | $CH_3-\overset{\overset{\displaystyle O}{\|\|}}{C}-CH_3$. |

## Section 20.10 *Carboxylic Acids, Esters, and Amides*

95.

| | Carboxylic Acid | IUPAC Name |
|---|---|---|
| (a) | $H-\overset{\overset{\displaystyle O}{\|\|}}{C}-OH$ | methanoic acid |
| (b) | $CH_3-\overset{\overset{\displaystyle O}{\|\|}}{C}-OH$ | ethanoic acid |
| (c) | $CH_3CH_2-\overset{\overset{\displaystyle O}{\|\|}}{C}-OH$ | propanoic acid |
| (d) | $CH_3CH_2CH_2-\overset{\overset{\displaystyle O}{\|\|}}{C}-OH$ | butanoic acid |

97.*

| | Ester | IUPAC Name |
|---|---|---|
| (a) | $H-\overset{\overset{\displaystyle O}{\|\|}}{C}-O-CH_2CH_3$ | ethyl methanoate |
| (b) | $CH_3-\overset{\overset{\displaystyle O}{\|\|}}{C}-O-CH_3$ | methyl ethanoate |

---

* continued on next page

97. Ester — IUPAC Name

(c) $CH_3CH_2-\overset{O}{\overset{\|}{C}}-O-CH_2CH_3$ — ethyl propanoate

(d) 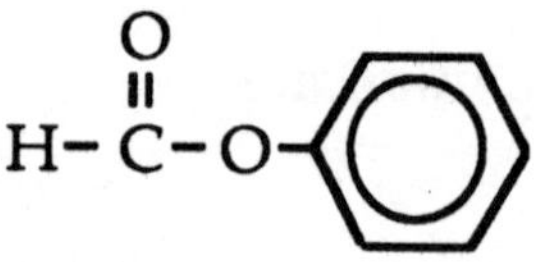 $H-\overset{O}{\overset{\|}{C}}-O-C_6H_5$ — phenyl methanoate

99. Amide — Name

(a) $H-\overset{O}{\overset{\|}{C}}-NH_2$ — methanamide (or formamide)

(b) $CH_3-\overset{O}{\overset{\|}{C}}-NH_2$ — ethanamide (or acetamide)

(c) $CH_3CH_2-\overset{O}{\overset{\|}{C}}-NH_2$ — propanamide (or propionamide)

(d) $CH_3CH_2CH_2-\overset{O}{\overset{\|}{C}}-NH_2$ — butanamide (or butyramide)

101. Esters cannot form hydrogen bonds.

103. The acid can form hydrogen bonds, the ester cannot.

105. $CH_3OH + [O] \rightarrow HCHO + H_2O$

107. (a) $CH_3\overset{O}{\overset{\|}{C}}-OH + CH_3OH \xrightarrow{H_2SO_4} CH_3\overset{O}{\overset{\|}{C}}-OCH_3 + H_2O$

(b) 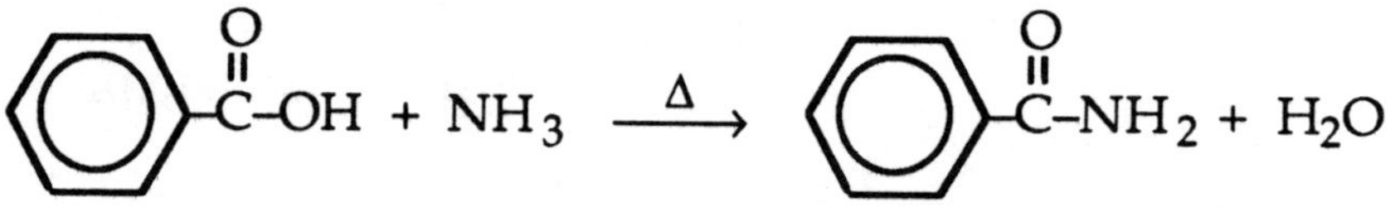 $C_6H_5-\overset{O}{\overset{\|}{C}}-OH + NH_3 \xrightarrow{\Delta} C_6H_5-\overset{O}{\overset{\|}{C}}-NH_2 + H_2O$

109. The ester phenyl ethanoate (or phenyl acetate) is produced from the reaction of phenol and acetic acid.

111. The amide phenyl ethanamide (or phenyl acetamide) is produced from the reaction of phenyl amine and acetic acid.

## General Exercises

113. The two reasons why an extraordinary number of organic compounds exist are:
   (1) carbon can self-link
   (2) organic molecules can rearrange to form isomers

115. Heptane

$CH_3CH_2CH_2CH_2CH_2CH_2CH_3$

Isooctane

$$\begin{array}{c} \quad\quad CH_3 \quad\quad\quad CH_3 \\ CH_3-\overset{|}{\underset{|}{C}}-CH_2-\overset{|}{CH}-CH_3 \\ CH_3 \quad\quad\quad\quad\quad\quad \end{array}$$

117.

| | Class | Solubility in Water |
|---|---|---|
| (a) | hydrocarbons | insoluble |
| (b) | organic halides | insoluble |
| (c) | alcohols | soluble |
| (d) | ethers | soluble |
| (e) | amines | soluble |
| (f) | amides | soluble |
| (g) | aldehydes | soluble |
| (h) | ketones | soluble |
| (i) | carboxylic acids | soluble |
| (j) | esters | insoluble |

119.

| | Compound | Classification |
|---|---|---|
| (a) | eicosane | saturated |
| (b) | limonene | unsaturated |
| (c) | cyclohexene | unsaturated |
| (d) | dodecane | saturated |

# LIST of ELEMENTS with their SYMBOLS and ATOMIC MASSES

| Name | Symbol | Atomic Number | Atomic Mass |
|---|---|---|---|
| actinium | Ac | 89 | (227) |
| aluminum | Al | 13 | 26.981539 |
| americium | Am | 95 | (243) |
| antimony | Sb | 51 | 121.75 |
| argon | Ar | 18 | 39.948 |
| arsenic | As | 33 | 74.92159 |
| astatine | At | 85 | (210) |
| barium | Ba | 56 | 137.327 |
| berkelium | Bk | 97 | (247) |
| beryllium | Be | 4 | 9.012182 |
| bismuth | Bi | 83 | 208.98037 |
| boron | B | 5 | 10.811 |
| bromine | Br | 35 | 79.904 |
| cadmium | Cd | 48 | 112.411 |
| calcium | Ca | 20 | 40.078 |
| californium | Cf | 98 | (251) |
| carbon | C | 6 | 12.011 |
| cerium | Ce | 58 | 140.115 |
| cesium | Cs | 55 | 132.90543 |
| chlorine | Cl | 17 | 35.4527 |
| chromium | Cr | 24 | 51.9961 |
| cobalt | Co | 27 | 58.93320 |
| copper | Cu | 29 | 63.546 |
| curium | Cm | 96 | (247) |
| dysprosium | Dy | 66 | 162.50 |
| einsteinium | Es | 99 | (252) |
| erbium | Er | 68 | 167.26 |
| europium | Eu | 63 | 151.965 |
| fermium | Fm | 100 | (257) |
| fluorine | F | 9 | 18.9984032 |
| francium | Fr | 87 | (223) |
| gadolinium | Gd | 64 | 157.25 |
| gallium | Ga | 31 | 69.723 |
| germanium | Ge | 32 | 72.61 |
| gold | Au | 79 | 196.96654 |
| hafnium | Hf | 72 | 178.49 |
| helium | He | 2 | 4.002602 |
| holmium | Ho | 67 | 164.93032 |
| hydrogen | H | 1 | 1.00794 |
| indium | In | 49 | 114.82 |
| iodine | I | 53 | 126.90447 |
| iridium | Ir | 77 | 192.22 |
| iron | Fe | 26 | 55.847 |
| krypton | Kr | 36 | 83.80 |
| lanthanum | La | 57 | 138.9055 |
| lawrencium | Lr | 103 | (260) |
| lead | Pb | 82 | 207.2 |
| lithium | Li | 3 | 6.941 |
| lutetium | Lu | 71 | 174.967 |
| magnesium | Mg | 12 | 24.3050 |
| manganese | Mn | 25 | 54.93805 |
| mendelevium | Md | 101 | (258) |
| mercury | Hg | 80 | 200.59 |
| molybdenum | Mo | 42 | 95.94 |
| neodymium | Nd | 60 | 144.24 |
| neon | Ne | 10 | 20.1797 |
| neptunium | Np | 93 | (237) |
| nickel | Ni | 28 | 58.69 |
| niobium | Nb | 41 | 92.90638 |
| nitrogen | N | 7 | 14.00674 |
| nobelium | No | 102 | (259) |
| osmium | Os | 76 | 190.2 |
| oxygen | O | 8 | 15.9994 |
| palladium | Pd | 46 | 106.42 |
| phosphorus | P | 15 | 30.973762 |
| platinum | Pt | 78 | 195.08 |
| plutonium | Pu | 94 | (244) |
| polonium | Po | 84 | (209) |
| potassium | K | 19 | 39.0983 |
| praseodymium | Pr | 59 | 140.90765 |
| promethium | Pm | 61 | (147) |
| protactinium | Pa | 91 | (231) |
| radium | Ra | 88 | (226) |
| radon | Rn | 86 | (222) |
| rhenium | Re | 75 | 186.207 |
| rhodium | Rh | 45 | 102.90550 |
| rubidium | Rb | 37 | 85.4678 |
| ruthenium | Ru | 44 | 101.07 |
| samarium | Sm | 62 | 150.36 |
| scandium | Sc | 21 | 44.955910 |
| selenium | Se | 34 | 78.96 |
| silicon | Si | 14 | 28.0855 |
| silver | Ag | 47 | 107.8682 |
| sodium | Na | 11 | 22.989768 |
| strontium | Sr | 38 | 87.62 |
| sulfur | S | 16 | 32.066 |
| tantalum | Ta | 73 | 180.9479 |
| technetium | Tc | 43 | (99) |
| tellurium | Te | 52 | 127.60 |
| terbium | Tb | 65 | 158.92534 |
| thallium | Tl | 81 | 204.3833 |
| thorium | Th | 90 | 232.0381 |
| thulium | Tm | 69 | 168.93421 |
| tin | Sn | 50 | 118.710 |
| titanium | Ti | 22 | 47.88 |
| tungsten | W | 74 | 183.85 |
| unnilennium | Une | 109 | (266) |
| unnilhexium | Unh | 106 | (263) |
| unniloctium | Uno | 108 | (265) |
| unnilpentium | Unp | 105 | (262) |
| unnilquadium | Unq | 104 | (261) |
| unnilseptium | Uns | 107 | (262) |
| uranium | U | 92 | 238.0289 |
| vanadium | V | 23 | 50.9415 |
| xenon | Xe | 54 | 131.29 |
| ytterbium | Yb | 70 | 173.04 |
| yttrium | Y | 39 | 88.90585 |
| zinc | Zn | 30 | 65.39 |
| zirconium | Zr | 40 | 91.224 |

# PERIODIC TABLE OF THE ELEMENTS

| | |
|---|---|
| ATOMIC NUMBER - | 1 |
| SYMBOL - | **H** |
| ATOMIC MASS - | 1.01 |

| | 1<br>IA | 2<br>IIA | 3<br>IIIB | 4<br>IVB | 5<br>VB | 6<br>VIB | 7<br>VIIB | 8<br>VIII | 9<br>VIII | 10<br>VIII | 11<br>IB | 12<br>IIB | 13<br>IIIA | 14<br>IVA | 15<br>VA | 16<br>VIA | 17<br>VIIA | 18<br>VIIIA |
|---|---|---|---|---|---|---|---|---|---|---|---|---|---|---|---|---|---|---|
| | | | | | | | | | | | | | | | | | | 2<br>**He**<br>4.00 |
| 2 | 3<br>**Li**<br>6.94 | 4<br>**Be**<br>9.01 | | | | | | | | | | | 5<br>**B**<br>10.81 | 6<br>**C**<br>12.01 | 7<br>**N**<br>14.01 | 8<br>**O**<br>16.00 | 9<br>**F**<br>19.00 | 10<br>**Ne**<br>20.18 |
| 3 | 11<br>**Na**<br>22.99 | 12<br>**Mg**<br>24.31 | | | | | | | | | | | 13<br>**Al**<br>26.98 | 14<br>**Si**<br>28.09 | 15<br>**P**<br>30.97 | 16<br>**S**<br>32.07 | 17<br>**Cl**<br>35.45 | 18<br>**Ar**<br>39.95 |
| 4 | 19<br>**K**<br>39.10 | 20<br>**Ca**<br>40.08 | 21<br>**Sc**<br>44.96 | 22<br>**Ti**<br>47.88 | 23<br>**V**<br>50.94 | 24<br>**Cr**<br>52.00 | 25<br>**Mn**<br>54.94 | 26<br>**Fe**<br>55.85 | 27<br>**Co**<br>58.93 | 28<br>**Ni**<br>58.69 | 29<br>**Cu**<br>63.55 | 30<br>**Zn**<br>65.39 | 31<br>**Ga**<br>69.72 | 32<br>**Ge**<br>72.61 | 33<br>**As**<br>74.92 | 34<br>**Se**<br>78.96 | 35<br>**Br**<br>79.90 | 36<br>**Kr**<br>83.80 |
| 5 | 37<br>**Rb**<br>85.47 | 38<br>**Sr**<br>87.62 | 39<br>**Y**<br>88.91 | 40<br>**Zr**<br>91.22 | 41<br>**Nb**<br>92.91 | 42<br>**Mo**<br>95.94 | 43<br>**Tc**<br>(99) | 44<br>**Ru**<br>101.07 | 45<br>**Rh**<br>102.91 | 46<br>**Pd**<br>106.42 | 47<br>**Ag**<br>107.87 | 48<br>**Cd**<br>112.41 | 49<br>**In**<br>114.82 | 50<br>**Sn**<br>118.71 | 51<br>**Sb**<br>121.75 | 52<br>**Te**<br>127.60 | 53<br>**I**<br>126.90 | 54<br>**Xe**<br>131.29 |
| 6 | 55<br>**Cs**<br>132.91 | 56<br>**Ba**<br>137.33 | 57<br>**La**<br>138.91 | 72<br>**Hf**<br>178.49 | 73<br>**Ta**<br>180.95 | 74<br>**W**<br>183.85 | 75<br>**Re**<br>186.21 | 76<br>**Os**<br>190.2 | 77<br>**Ir**<br>192.22 | 78<br>**Pt**<br>195.08 | 79<br>**Au**<br>196.97 | 80<br>**Hg**<br>200.59 | 81<br>**Tl**<br>204.38 | 82<br>**Pb**<br>207.2 | 83<br>**Bi**<br>208.98 | 84<br>**Po**<br>(209) | 85<br>**At**<br>(210) | 86<br>**Rn**<br>(222) |
| 7 | 87<br>**Fr**<br>(223) | 88<br>**Ra**<br>(226) | 89<br>**Ac**<br>(227) | 104<br>**Unq**<br>(261) | 105<br>**Unp**<br>(262) | 106<br>**Unh**<br>(263) | 107<br>**Uns**<br>(262) | 108<br>**Uno**<br>(265) | 109<br>**Une**<br>(266) | | | | | | | | | |

| | | | | | | | | | | | | | |
|---|---|---|---|---|---|---|---|---|---|---|---|---|---|
| 58<br>**Ce**<br>140.12 | 59<br>**Pr**<br>140.91 | 60<br>**Nd**<br>144.24 | 61<br>**Pm**<br>(147) | 62<br>**Sm**<br>150.36 | 63<br>**Eu**<br>151.97 | 64<br>**Gd**<br>157.25 | 65<br>**Tb**<br>158.93 | 66<br>**Dy**<br>162.50 | 67<br>**Ho**<br>164.93 | 68<br>**Er**<br>167.26 | 69<br>**Tm**<br>168.93 | 70<br>**Yb**<br>173.04 | 71<br>**Lu**<br>174.97 |
| 90<br>**Th**<br>232.04 | 91<br>**Pa**<br>(231) | 92<br>**U**<br>238.03 | 93<br>**Np**<br>(237) | 94<br>**Pu**<br>(244) | 95<br>**Am**<br>(243) | 96<br>**Cm**<br>(247) | 97<br>**Bk**<br>(247) | 98<br>**Cf**<br>(251) | 99<br>**Es**<br>(252) | 100<br>**Fm**<br>(257) | 101<br>**Md**<br>(258) | 102<br>**No**<br>(259) | 103<br>**Lr**<br>(260) |